Audio Production Principles

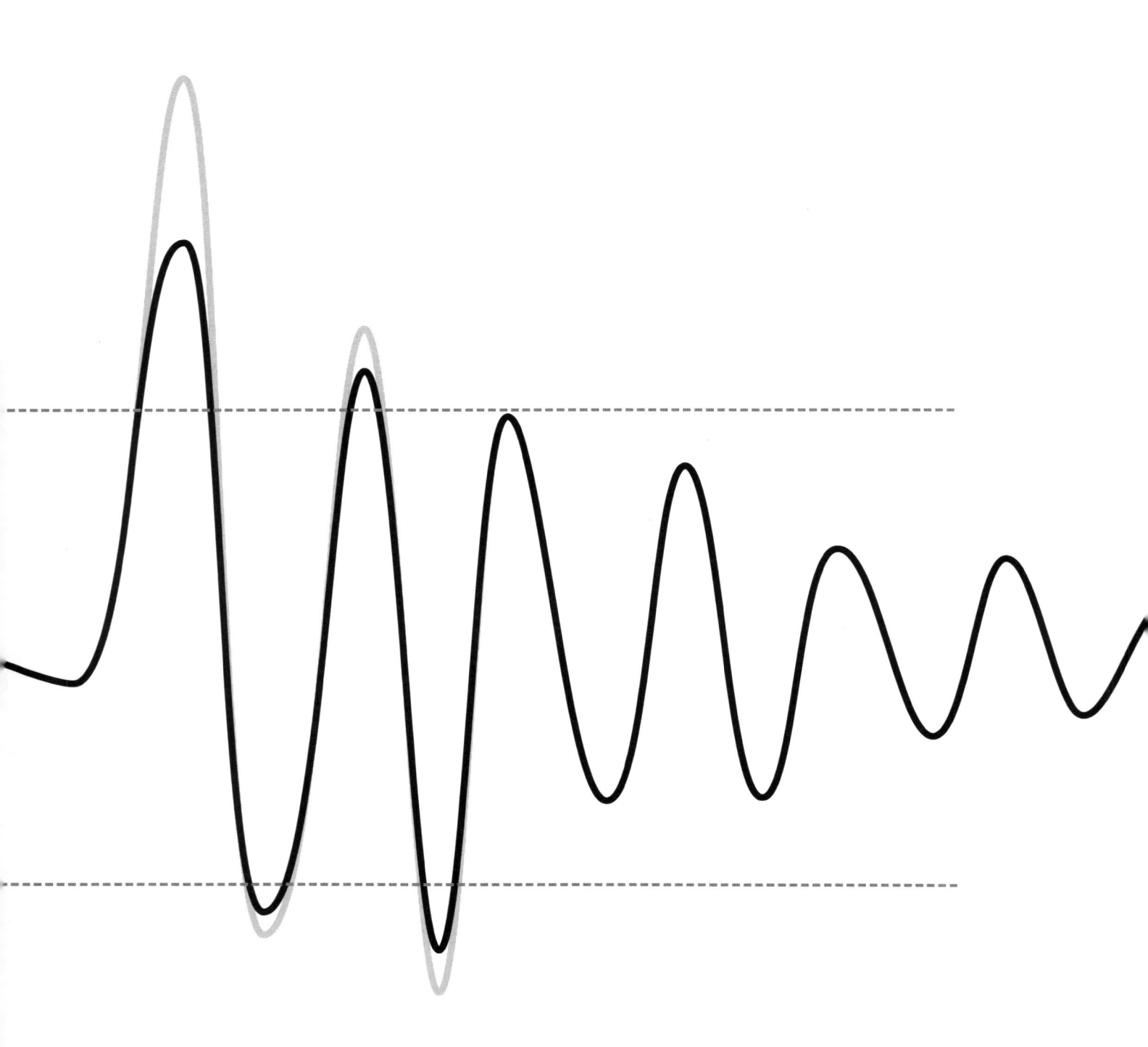

Audio Production Principles

Practical Studio Applications

Stephane Elmosnino

OXFORD
UNIVERSITY PRESS

OXFORD
UNIVERSITY PRESS

Oxford University Press is a department of the University of Oxford. It furthers
the University's objective of excellence in research, scholarship, and education
by publishing worldwide. Oxford is a registered trade mark of Oxford University
Press in the UK and certain other countries.

Published in the United States of America by Oxford University Press
198 Madison Avenue, New York, NY 10016, United States of America.

© Oxford University Press 2018

All rights reserved. No part of this publication may be reproduced, stored in
a retrieval system, or transmitted, in any form or by any means, without the
prior permission in writing of Oxford University Press, or as expressly permitted
by law, by license, or under terms agreed with the appropriate reproduction
rights organization. Inquiries concerning reproduction outside the scope of the
above should be sent to the Rights Department, Oxford University Press, at the
address above.

You must not circulate this work in any other form
and you must impose this same condition on any acquirer.

Library of Congress Cataloging-in-Publication Data
Names: Elmosnino, Stephane, author.
Title: Audio Production Principles : Practical Studio Applications / Stephane Elmosnino.
Description: New York, NY : Oxford University Press, [2018] | Includes index.
Identifiers: LCCN 2017017674 | ISBN 9780190699369 (pbk. : alk. paper) |
ISBN 9780190699352 (cloth : alk. paper) | ISBN 9780190699390 (companion website) |
ISBN 9780190699376 (updf) | ISBN 9780190699383 (epub)
Subjects: LCSH: Sound–Recording and reproducing. | Sound studios–Equipment and supplies.
Classification: LCC TK7881.4 .E567 2018 | DDC 781.49–dc23
LC record available at https://lccn.loc.gov/2017017674

9 8 7 6 5 4 3 2 1
Paperback printed by WebCom, Inc., Canada
Hardback printed by Bridgeport National Bindery, Inc., United States of America

To my family

Contents

Acknowledgments • xiii

About the Companion Website • xv

1. **Introduction** • 1
    1.1. Critical Listening • 2
2. **Studio Design Basics** • 5
    2.1. Mixing Room: Standing Waves • 6
    2.2. Mixing Room: Symmetry • 7
    2.3. Mixing Room: Optimizations • 7
3. **Preproduction** • 20
    3.1. DAW Session Preparation • 22
    3.2. Tempo • 23
    3.3. Structure • 24
    3.4. Guide Tracks • 24
    3.5. Sound "Map" • 25
    3.6. Studio Hierarchy • 25
    3.7. Budget • 26
4. **Microphone Basics** • 28
    4.1. Choosing the Right Microphone • 29
    4.2. Microphone Types and Attributes • 29
    4.3. Polar Patterns • 30
    4.4. Physical Size • 31
    4.5. Distance • 32
    4.6. Orientation • 33
    4.7. Other Considerations • 34
    4.8. Microphone Selection Recap • 34
    4.9. Stereo Recording Techniques • 35
    4.10. Stereo Recording Technique and Polar Pattern Selection Recap • 40
    4.11. Creating Interest with Microphones • 42
5. **Recording** • 44
    5.1. Recording Medium • 45
    5.2. Gain Staging • 45

- 5.3. Bit Depth • **45**
- 5.4. Sample Rate • **47**
- 5.5. Aliasing • **47**
- 5.6. Recording Live versus Overdubbing • **48**
- 5.7. Recording Order • **51**
- 5.8. Mix While Recording • **52**
- 5.9. Running the Session • **52**
- 5.10. Headphone Mix • **54**
- 5.11. Pulling the Right Sound • **55**

6. Drum Kit • 57
- 6.1. The Room • **58**
- 6.2. The Instrument • **58**
- 6.3. The Microphones • **60**
- 6.4. The Overheads • **64**
- 6.5. Running the Session • **70**
- 6.6. Editing • **71**
- 6.7. Mixing Tips • **73**
- 6.8. Creating Interest with Drums • **79**

7. Bass Guitar • 81
- 7.1. The Room • **82**
- 7.2. The Instrument • **82**
- 7.3. The Microphones • **82**
- 7.4. The Processors • **83**
- 7.5. Running the Session • **84**
- 7.6. Mixing Tips • **84**
- 7.7. Creating Interest with Bass • **87**

8. Electric Guitar • 88
- 8.1. The Room • **89**
- 8.2. The Instrument • **89**
- 8.3. The Microphones • **93**
- 8.4. Pulling the Right Sound • **95**
- 8.5. Running the Session • **96**
- 8.6. Creating Interest with Electric Guitars • **97**

9. Acoustic Guitar • 98
- 9.1. The Room • **99**
- 9.2. The Instrument • **99**
- 9.3. The Microphones • **99**
- 9.4. The Processors • **103**
- 9.5. Creating Interest with Acoustic Guitars • **104**

10. Vocals • 105
- 10.1. The Room • 106
- 10.2. The Instrument • 106
- 10.3. The Microphones • 107
- 10.4. The Processors • 110
- 10.5. Running the Session • 111
- 10.6. Editing • 114
- 10.7. Tuning • 116
- 10.8. Mixing Tips • 116

11. Synthesis Basics • 120
- 11.1. Waveforms • 121
- 11.2. Filters • 123
- 11.3. Modulators • 124
- 11.4. Frequency-Modulation Synthesis • 128
- 11.5. Amplitude and Ring Modulation • 128
- 11.6. Instrument Design • 129
- 11.7. Enhancing Recorded Instruments • 129
- 11.8. Programming Real Instruments • 129

12. Mixing • 131
- 12.1. The Mix as a Virtual Soundstage • 132
- 12.2. Mixing Workflows • 133
- 12.3. Monitoring Level • 146
- 12.4. Reference Tracks • 147
- 12.5. Different Perspective • 147
- 12.6. Target Audience and Intended Playback System • 147
- 12.7. Mix-Bus Processing • 148
- 12.8. Hardware Emulations • 149
- 12.9. 1001 Plug-ins • 149
- 12.10. Tactile Control • 149
- 12.11. Hot Corners • 149
- 12.12. When a Part Will Not Fit • 150

13. Panning • 151
- 13.1. Intensity versus Phase Stereo • 151
- 13.2. Pan Law • 154
- 13.3. Panning Tips • 155

14. Equalizers • 161
- 14.1. Fixing Equalizers • 162
- 14.2. Separation Equalizers • 162
- 14.3. Enhancement Equalizers • 163

14.4. Q Setting • 163
14.5. Filter Type • 164
14.6. Linear-Phase Equalizers • 165
14.7. Active and Passive Equalizers • 167
14.8. Equalizer Tips • 167
14.9. Plug-in Presets and Graphic Equalizers • 173

15. Compression, Expansion, and Gates • 174
15.1. Hardware Designs • 175
15.2. Compressor Settings • 179
15.3. Why Use Compressors? • 185
15.4. Setting Up a Compressor • 187
15.5. Parallel Compression • 188
15.6. Serial Compression • 189
15.7. Equalizer-Compressor-Equalizer • 190
15.8. Increasing Reverb • 190
15.9. Side-Chain Input Uses • 190
15.10. De-Essers • 199
15.11. Multiband Compressors • 200
15.12. Low-, Mid-, and High-Level Compression • 202
15.13. Negative Ratios • 203
15.14. Expanders and Gates • 203
15.15. Spill Reduction • 205
15.16. Using Gates and Expanders in a Mix • 206
15.17. Upward versus Downward • 206

16. Reverberation and Delays • 208
16.1. Reverb Type • 209
16.2. Early Reflections/Reverb Tail • 212
16.3. Placement Reverb • 213
16.4. Enhancement Reverb • 214
16.5. Reverb for Depth • 215
16.6. Reverb Settings • 216
16.7. Using Convolution Reverbs • 217
16.8. Reverb Tips • 217
16.9. Delays • 221

17. Saturation • 225
17.1. Hard Clipping • 226
17.2. Soft Clipping • 226
17.3. Limiters • 226
17.4. Summing Console Emulation • 226
17.5. Analog Modeled Plug-ins • 227

- 17.6. Tape Saturation • **227**
- 17.7. Tube Saturation • **227**
- 17.8. Harmonic Enhancers • **227**
- 17.9. Parallel Processing • **228**

18. Mastering • **229**
- 18.1. Delivery Format • **230**
- 18.2. Monitoring Environment • **231**
- 18.3. Metering • **231**
- 18.4. Room Tone • **232**
- 18.5. Track Order • **233**
- 18.6. Fade Out and Pause • **234**
- 18.7. Mastering Your Own Mix • **234**
- 18.8. Sending a Mix to the Mastering Engineer • **235**
- 18.9. Mastering Tools • **236**
- 18.10. Deliverables • **246**
- 18.11. MP3 • **247**

19. Conclusion • **249**

Index • **251**

Acknowledgments

Many people need to be thanked for the completion of this book. Firstly, thank you to the team at Oxford University Press, particularly to my editor, Norman Hirschy, for believing in my work and helping me develop this project as much as he did.

Thank you to the amazing musicians and friends who have offered their talent to record the audio examples in this book: Thomas Combes, the best guitarist I know, Daniel Hitzke for the superb drumming, and Daphné Maresca for the amazing vocals.

My deepest thanks to my family for your continued support over the years.

Thank you to all of those who have inspired me over the years: the musicians I have had the pleasure of working with, my various mentors who have freely shared their knowledge, the lecturers I have had the chance of teaching alongside, and my sister Davina for inspiring me by her incredible musical production talent. And finally, thank you to my students; you have taught me more than you can imagine.

About the Companion Website

www.oup.com/us/audioproductionprinciples
Username: Music2
Password: Book4416

Oxford has created a website to accompany *Audio Production Principles*. Materials that cannot be made available in a book, namely, audio examples, are provided here. The reader is encouraged to consult this resource in conjunction with the book. Examples available online are indicated in the text with Oxford's symbol ⏵. A high-quality monitoring environment is crucial in order to hear the finer differences between audio examples.

Chapter 1

Introduction

A question that I often get asked by audio engineering students is "What is your opinion on tool X/Y/Z?" While the answer may feel as though I were diverting from the question, it is quite simple. Audio engineering tools are just that: tools. Just as a builder uses hammers and nails, audio engineers use equalizers (EQs), compressors, and reverberators (more commonly referred to as "reverbs"). When you need to put a nail in a plank of wood, a hammer is definitely the best tool for the job, but other options may also work. The same concept applies to audio engineering: some tools are better suited to certain tasks than others. Which tool ends up being used for each particular task will largely depend on knowledge, experience, and availability of the best tool for the job. One of the aims of this book is to look at the tools available to producers and engineers to allow you to get a head start when deciding which tools and techniques may be the most appropriate starting point in a given situation. Note that no amount of information or numbers of tips, tricks, and tools will ever replace experience. While it is true that more knowledge will always help in your quest to perfect recordings, having the ear for what sounds good and what needs more work still remains the most important factor in whether a production sounds professional or not. It has often been said that a good balance of experience, knowledge, and equipment is necessary in order to sustain a career as an audio engineer.

Another concept that must be mentioned at the start of this book is to get the sound "right" at the source. The most common mistake that audio students and amateur engineers make is to rush through the early stages of the recording project with a "we'll fix it in the mix" attitude. While the access to high-end post-processing tools allows for some flexibility, a clear idea of how the instruments will sound and where the project is headed is necessary. The time spent on a particular song or project will be fairly similar, whether it was thirty minutes for moving microphones at the recording stage or equalizing instruments during the mixing session. The only difference is that better results are always achieved when the time is spent early in the project. Similarly, if everything in the mix has been fixed in terms of EQ and dynamics, for example, working with enhancement tools should be a quick process. If the mix sounds too compressed, you have probably used compressors too early and should have worked on volume automation first. If the mix sounds drowned in reverb, you have probably set up your reverbs before equalizing and balancing the mix. If the mix sounds muddy in the low end, you have probably made EQ boosts before doing cuts and adjusting levels. Spending time getting the source right will always yield better results.

The last point to remember is to always question what you read, watch, hear, and discuss. Different engineers work differently, and what works for one person may be completely useless for another. Even the processes and guidelines outlined in this book are meant to be questioned and used in your own way!

1.1 Critical Listening

The concept of critical listening and the need for audio engineers to develop their aural skills is not a new one. It is crucial for you to learn from the work of others and be able to judge sound quality in order to become a better engineer. To that end, the process of "reverse engineering" commercial mixes plays a very important part in developing your ear. The ability to zoom in on individual instruments within a mix and accurately judge their sonic qualities is a skill that builds over multiple years of practice. It is also necessary for audio engineers while judging sound quality to recognize the tools that may have been used to achieve a particular sound. For example, if the voice in a recording is bass heavy, has natural reverb and ambience, and contains very little sibilance, you need to be able to suggest that the low end could not have been achieved by using proximity effect of the microphone, because of the lack of sibilance and presence of natural reverb. This skill can then be transferred in the studio when you are recording or mixing to achieve the desired outcome. While subjective terms such as "thick," "punchy," and "boxy" are often used as descriptors, it is important to recognize the tools that may have been used to create a particular sound.

In order to judge the sonic qualities of a sound, you should take a few elements into consideration:

- Spectral content: Which area contains the most power? Has this content been created naturally or modified through the use of EQ? Does the sound contain many harmonics and overtones? Are they created naturally or through saturation processes? Does the sound contain noise (either as an integral part of the intended sound, or extra noise added during the recording process)?
- Dynamic envelope: How does the overall loudness evolve over time? Has it been modified through audio processing? How does each frequency decay over time? Are the different decaying frequency times created naturally or through post-processing?
- Spatial attributes: Which space has the sound been recorded in? What is the decay time of each frequency area? Where in the stereo field is the sound placed and by using which panning method?

There are many commercial resources available for further developing your critical listening skills. The more theoretical books on this topic include William Moylan's *Understanding and Crafting the Mix: The Art of Recording* (Burlington: Focal Press, 2002) and Jason Corey's *Audio Production and Critical Listening: Technical Ear Training* (Burlington: Focal Press, 2010). These books also offer exercises and methods for developing critical listening skills. From a more practical standpoint, other resources are available, such as F. Alton Everest's *Critical Listening Skills for Audio Professionals* (Boston: Thomson Course Technology, 2007) and the many stand-alone and web-based applications such as "Quiztones."

Developing your engineering skills is something that you can do every day. Keep learning by actively listening to music and sounds. Listen for production tricks and sonic characteristics; think about the listening environment and how it affects the production. For example, if you hear a song in a shopping center, how does the mix enhance your experience of the song in this environment? Which instruments are most prominent? Which ones are buried in the mix? How is the playback volume affecting your perception of the mix? How is the speaker coloring the mix? Audio engineering is a craft that requires constant developing and refining. No one becomes a great engineer overnight, and even the greats of this world keep learning new techniques, refining old ones, and developing new understandings constantly.

Chapter 2

Studio Design Basics

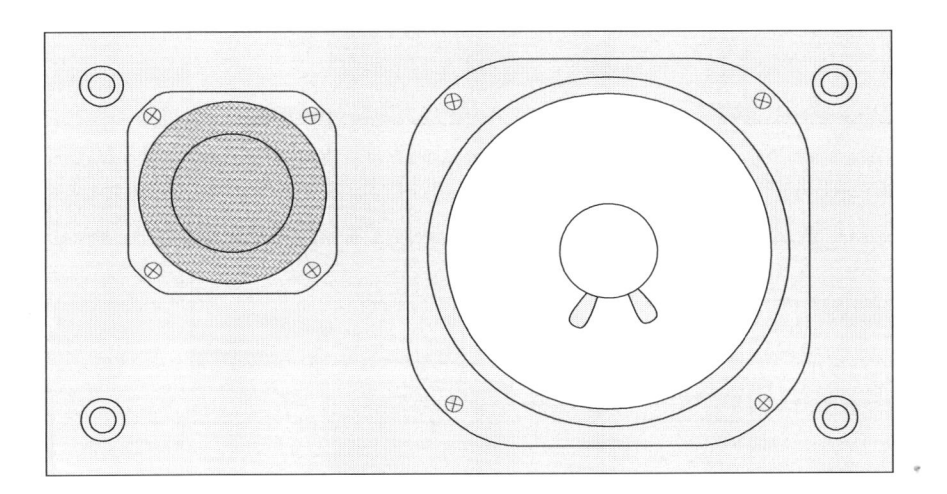

Monitors and the room in which they are placed are the most important tools of the studio engineer. It is important to understand that the monitoring environment is a "system": all the different elements of this monitoring system interact with one another. Within this system, the size, shape, and construction materials of the room are almost as important as the quality of the monitors and their placement.

In the early stage of setting up a studio, a limited budget often means that sacrifices must be made. It is advisable to start by getting different monitors that cater for areas in which the main monitors lack. For example, adding cheap computer speakers can help a great deal in getting bass and vocals volume right. Since computer speakers are designed to enhance speech frequencies, if the vocals are too low in the mix on those frequencies, there is a good chance that they are too low overall. Because computer speakers will also distort long before the threshold of pain is reached, they can be pushed hard to see how much low end is contributing to the distortion. Having different pairs of headphones can also help, especially when you are mixing in an untreated room. Some headphones excel in high-end detail, and some deal better with the low end; you should research and try different pairs before settling on those that best complement each other.

Another addition to the monitoring system that should be considered is a subwoofer. Modern music requires most studios to work with one unless full-range speakers are used. One of the main advantages of using a subwoofer is that you are able to monitor with extended low frequencies even at low listening levels. Human hearing is not linear in terms of its frequency response, and without a subwoofer, we need higher volume levels to hear low frequencies. Another main advantage of using a subwoofer is that electronic/club-music mixes translate better to larger systems. Finally, when you use a subwoofer with crossover, the main monitors may run more cleanly, as the "heavy lifting" in the low end is redistributed to the subwoofer. The good news is that expensive subwoofers are not necessarily essential. In fact, a cheap subwoofer can sound great if properly used. Since subwoofers represent a narrow frequency band, using cheap high-fidelity units may be good enough as long as they are not pushed hard and are properly positioned, low-pass filtered, and at the correct volume. Hi-fi subwoofers running at less than half of their capacity can do an excellent job in a music studio environment.

2.1 Mixing Room: Standing Waves

One of the main issues that you must deal with when you are designing a room is standing waves: low frequencies (below about 300 Hz) that are reflected between two parallel walls. As these frequencies get reflected in the room, their level rises or lowers though phase reinforcement or cancellation. While standing waves are unavoidable, you can take steps in order to minimize the issues they cause. The first step is to ensure that the mixing room's height, width, and length are all different. This variation reduces the probability that the standing waves created by the different room boundaries are all different, and therefore not reinforcing the problem. In some cases, having an angled back wall and/or ceiling can help reduce the standing waves being created to the side walls only. Note that if the ceiling is angled, the speakers should be placed on the short wall, facing the "open side" of the room (Figure 2.1). This placement ensures that the reflected frequencies on the ceiling are "sent" to the back of the room instead of reflected back to the listening position.

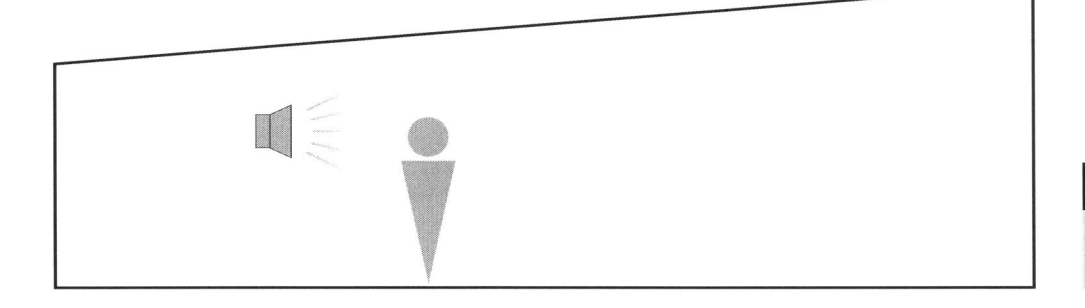

FIGURE 2.1
Speaker placement in room with angled ceiling.

2.2 Mixing Room: Symmetry

While symmetry should be avoided in recording rooms, mixing rooms require quite the opposite. Ensuring that the room is symmetrical in terms of speaker and rack equipment placement can help a lot in achieving better stereo imaging. Should the room have non-symmetrical features such as doors, windows, and wall construction materials, these features should be kept behind the listening position. Everything in front of the listening position should be as identical as possible on either side of the room. A good way to check whether your room is symmetrical enough is to run white or pink noise through each monitor one at a time, and listen for differences in tone when you are switching from one to the other.

2.3 Mixing Room: Optimizations

Optimizing the sound in the control room starts long before any sort of acoustic treatment is put up. Placing the speakers in the right position is the most important type of optimization that you can do. Poorly placed speakers in a fully treated room will sound worse than well-placed speakers in an untreated room. As a rule of thumb, two-thirds of the optimization comes from speaker placement, one-third is acoustic treatment, and the "icing on the cake" is room-correction software. These three elements are discussed in depth over the next few pages.

2.3.1 Monitors

2.3.1.1 Main Monitors Placement
The main monitors placement is critical in obtaining the best sound from the system. There are two main issues, which are addressed at this point: minimizing

speaker/boundary interactions and minimizing the room's modal problems. The first decision to be made relates to the position of the main monitors, depending on the overall shape of the room. Because most mixing rooms are rectangular, it is often preferable to place the speakers facing the long side. In this case, a good starting point is to position speakers at 70% of the room width from the front wall (Figure 2.2).

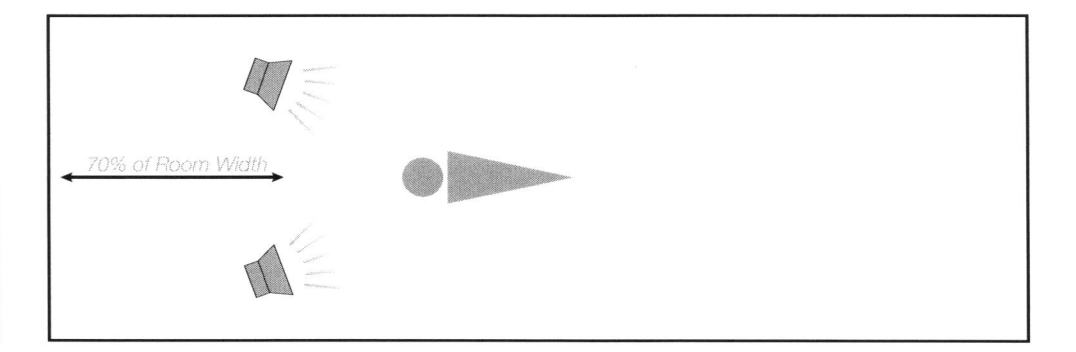

FIGURE 2.2
Speaker placement: distance from front wall.

Ensuring that the monitors are at least 60 cm away from any wall is recommended in order to minimize speaker/boundary interactions. As speakers are placed closer to the walls, there is a buildup of bass that tends to muddy the low-end response of the monitors. You could, however, ignore this advice in the event that the monitors are too small and require a low-end boost! If you are using main monitors with a frequency response that extends in the sub-bass, avoid placing them between 1 m and 2 m from the front wall.

It is also advisable to have the listening position at 38% of the room length from the front wall (Figure 2.3). This placement helps ensure that room modes for the length are minimized.

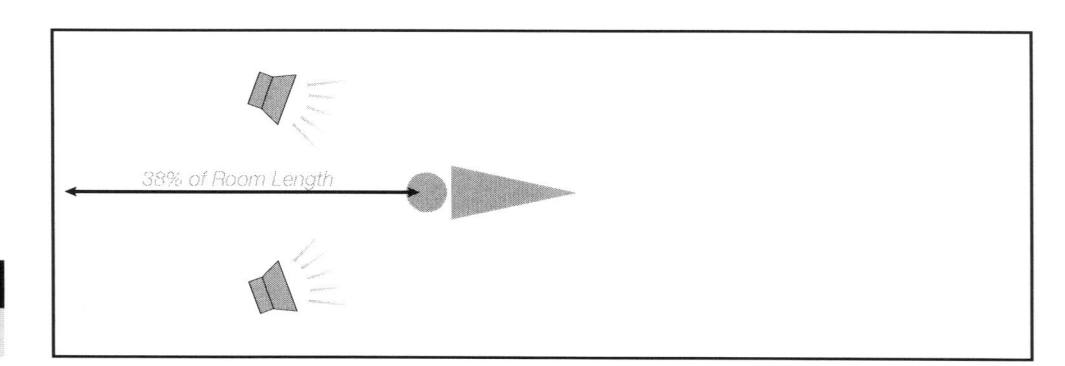

FIGURE 2.3
Ideal listening position.

When you are positioning the speakers in relation to the listening position, the rule of thumb is to create an equilateral triangle between the speakers and the head of the listener (Figure 2.4). The speakers can then be placed closer to each other if there is a hole in the stereo image created during playback, or farther away if the playback seems too monophonic. While it is important to ensure that the tweeters are facing the listener's ears as a starting point, the "sweet spot" created by such configuration often extends to the inside of the triangle. Because of this effect, many engineers prefer to sit slightly forward to ensure that the largest part of the sweet spot is being used.

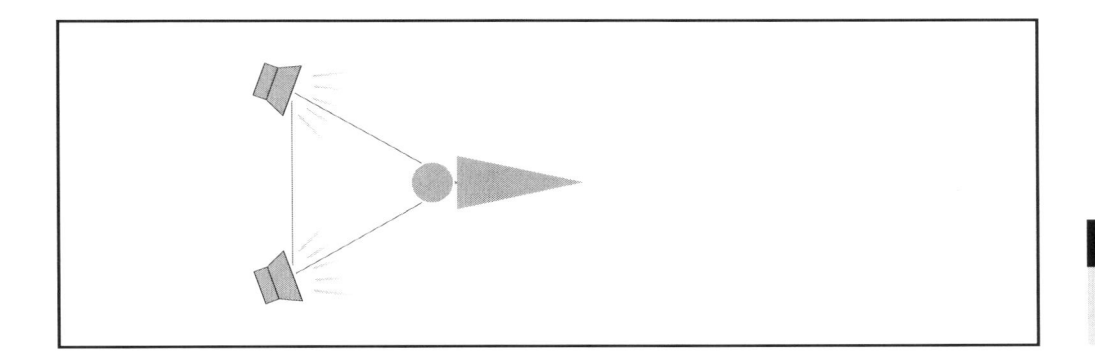

FIGURE 2.4
Speaker placement in relation to listening position.

If at all possible, slightly angling the main monitors up or down while they are still facing the listener's head can help reduce the room's modal problems. Note that too much angle can reduce the size of the sweet spot, as moving the listening position forward and back will change the ears' position in relation to the tweeters.

If the speakers are behind computer screens, make sure that the direct sound is not being blocked. If the screens must be placed on the same axis as the speakers from the listening position, lowering them can work well in keeping the direct sound free from obstacles (Figure 2.5).

FIGURE 2.5
Monitor placement in relation to speakers.

2.3.1.2 Subwoofer Placement

For modern genres of music, using a subwoofer is recommended for fully hearing the frequencies present in the low end. It is, however, essential to find an adequate placement for the subwoofer, as it can cause more issues than it solves if not properly placed. Unless you are prepared to spend hours in testing the system and optimizing the subwoofer placement, do not use one!

While correct subwoofer placement requires the use of room-analysis software, you can do a fair amount of optimization by ear. To begin with, set the subwoofer at the mixing position and walk around the room while playing bass-heavy music, ensuring that all the notes in the lower octave of the spectrum are being played. The bass should "come alive" in some areas of the room; take note of them. Since bass is omnidirectional, the subwoofer can then be placed on the noted positions and should sound similar at the listening position. A good starting point to listen to is along an imaginary circle that runs through both speakers, with its center being the listening position (Figure 2.6). On this circle, the best position is often between the main monitors, slightly to the left or right (which often equates to being roughly at one-third of the room width).

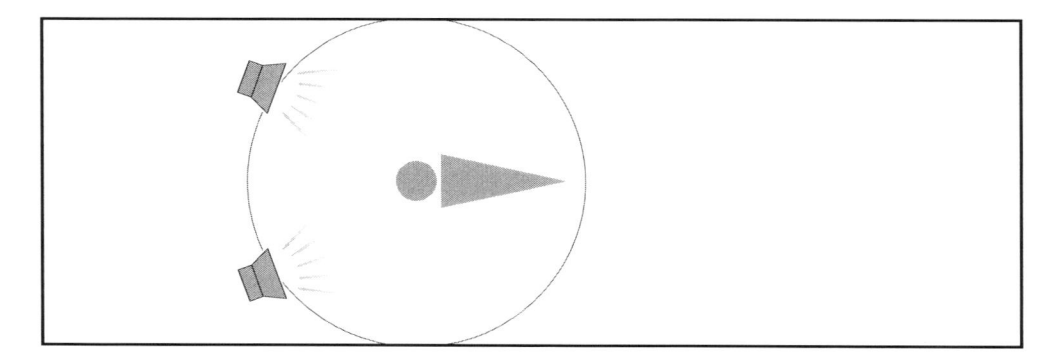

FIGURE 2.6

Subwoofer placement: starting point.

Chances are that a few different positions along that circle will yield the best results, which you can then further experiment with later by using room-analysis software. From there, you should spend many hours of recording room response to the different placements to determine the position that gives the flattest low-end frequency response. The subwoofer will sound different, depending on its position within the room, its orientation (facing the listening position is not necessarily the best), and its height. While subwoofers are often placed on the ground and against a wall to mimic the sound's behavior if they were flush mounted, lifting them to head level can do two things to clean up the low end: reduce subwoofer/ground interactions (similar to how guitar amplifiers are placed on chairs to avoid muddying the sound), and give more direct rather than reflected sound.

Some subwoofers have the option of altering the phase of the signal being reproduced either as a switch inverting the polarity, or as a continuous phase knob. When experimenting with these settings, try to listen to where the subwoofer frequencies come from. There should be an illusion that the low end is coming from the main monitors, not the subwoofer. Note that in no circumstance should the low end feel as if it were coming from the subwoofer. If it does, the low-pass filter on the subwoofer may be set too high, or its amp might be driven too hard, giving venting noise or harmonic distortion.

Large mixing rooms can benefit from using two subwoofers. The main benefit of this use is to give the impression that sub-bass is surrounding the listener. You can also use a second subwoofer to lighten up the load on the first subwoofer by running both at a lower volume. If you are using two subwoofers, set one up first with the help of all the techniques described in this chapter, then repeat them for the second one. Since sub-bass is omnidirectional, symmetry in subwoofer placement is not necessary. In most cases, placing the second subwoofer between the main monitors is a good starting point. If the first subwoofer was placed in this area, the second will most likely need to be placed symmetrically along the width of the room, and less than 1.5 m away from the first subwoofer.

2.3.1.3 Subwoofer Crossover

The decision of using a crossover to separate the signal going to the subwoofer from the signal going to the main monitors or running the full signal to the main monitors and only low-pass-filtering the subwoofer can be tricky. Each setup has pros and cons, which are discussed here.

If the main monitors are very small and do not reproduce low end accurately (below roughly 80 Hz), no crossover is necessary. In this case, running the full signal to the main monitors and low-pass-filtering the subwoofer is preferable. The reason for not using a crossover in this case is that it would introduce unnecessary electronics in the system given that the main monitors are unable to reproduce low frequencies and therefore would not introduce phase issues between the subwoofer and the mains. The exception can be made if using a high-quality crossover, which could help lighten the low-end-frequency load on the main monitors, therefore cleaning up their midrange-frequencies. When you are using main monitors that extend well into the low frequencies, not using a crossover can have the issue of creating low-end phasing between the different speakers and the subwoofer. This issue can, however, be corrected if you properly place the subwoofer by using test equipment. In this case, the low-pass filter of the subwoofer should be set quite low to ensure that it is used only to give support to the main monitors. Working this way allows you to use lower-quality subwoofers, as they are not being driven hard and do not have to reproduce many frequencies. In some cases, cheap hi-fi-system subwoofers can be used in the studio when set this way. The low-pass filters included in such subwoofers are often low in quality. It is therefore recommended that you use an extra low-pass filter to ensure that only the lowest frequencies are run through the subwoofer. In the case of large main monitors, using a crossover would have some midrange-frequency benefits such as adding dynamic range and depth. Note that the smaller the main monitors, the higher the crossover frequency should be (with a maximum around 80 Hz). Another important point is that you should use the crossover built into the subwoofer only when you are working with high-quality equipment, as lower-quality crossovers can negatively affect the signal going to the main monitors.

2.3.1.4 *Other Considerations*

Monitors' amplification should be able to run much higher than what is needed to ensure that they run cleanly all the time. This precaution is especially true when you are dealing with transient heavy sounds that may otherwise distort for a very short period of time (which may be too short to be heard as distortion but will blur the sounds and make mixing harder).

When choosing speakers, listen for depth and how they produce a 3D image of the music being listened to. Stereo imaging is very dependent on the room in which speakers are placed, and most manufacturers produce speakers that are flat across the frequency range (if they are placed in an anechoic chamber). This means that the main difference between monitors is their ability to convey a sense of depth. The size of the sweet spot they can create is also dependent on monitor design, but this dependence is nearly impossible to judge in a showroom. Overall, choosing monitors that create a great 3D image will affect how much compression is used on mixes, reverberation, and other depth-related processors.

2.3.2 Acoustic Treatment

Once you have found the best position for the main monitors and subwoofer, acoustic treatment can help further enhance the monitoring environment first by solving reflections to the listening position, then reducing room-modal problems, and finally adjusting the room-size perception. For a rule of thumb, you should spend the same amount of money on room treatment as you spend on monitors. A great monitoring environment will make mixes sound a lot more "expensive" than any fancy equalizer or compressor. No number of plug-ins or hardware units will help if the finer details of the mix cannot be properly heard.

2.3.2.1 *Reflection-Free Zone*

The first problem that you should fix in an untreated room is the first reflections to the listening position (Figures 2.7 and 2.8). The aim is to create a reflection-free zone (RFZ) at the listening position (Figures 2.9 and 2.10). The RFZ ensures that the first reflections from the speakers, which are the loudest after the direct sound, are either absorbed or diffused. In a rectangular room, there are ten first reflections to deal with (two for each wall and the ceiling). An easy way to figure out where to place the acoustic treatment is to sit at the listening position and have someone slide a mirror along each wall until the monitors can be seen. Since mid- and high frequencies are reflected similarly to light (at an angle equal to the one hitting a boundary), the sound leaving the monitors will hit each wall where the mirror shows the speakers and be reflected to the listening position.

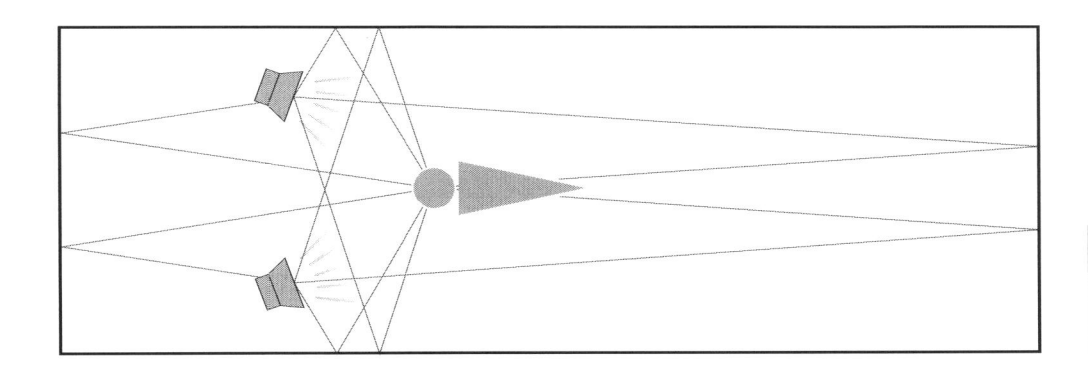

FIGURE 2.7
First reflections at the listening position: top view.

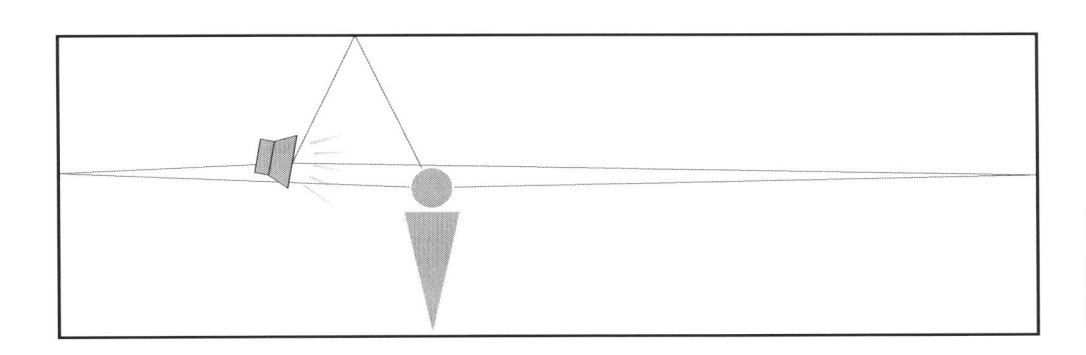

FIGURE 2.8
First reflections at the listening position: side view.

Note that the "least important" wall to treat for first reflections is the front wall behind the speakers. As high frequencies are more directional, only mid-frequencies need to be dealt with on this wall, a treatment often done by using broadband bass traps on this wall. If you are using absorption panels to create the RFZ, it is a good idea to angle acoustic panels away from the listening position at the first-reflection points on the side walls and ceiling to both absorb and reflect at a different angle (Figures 2.9 and 2.10).

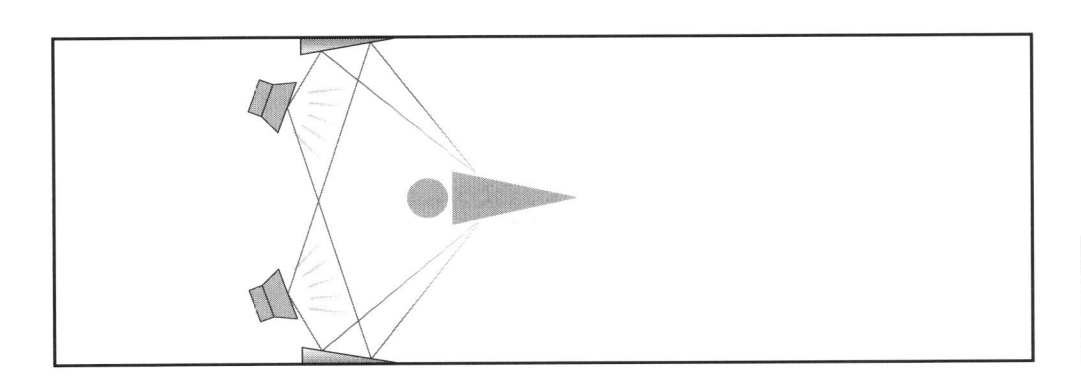

FIGURE 2.9
Creating a reflection-free zone: top view.

FIGURE 2.10
Creating a reflection-free zone: side view.

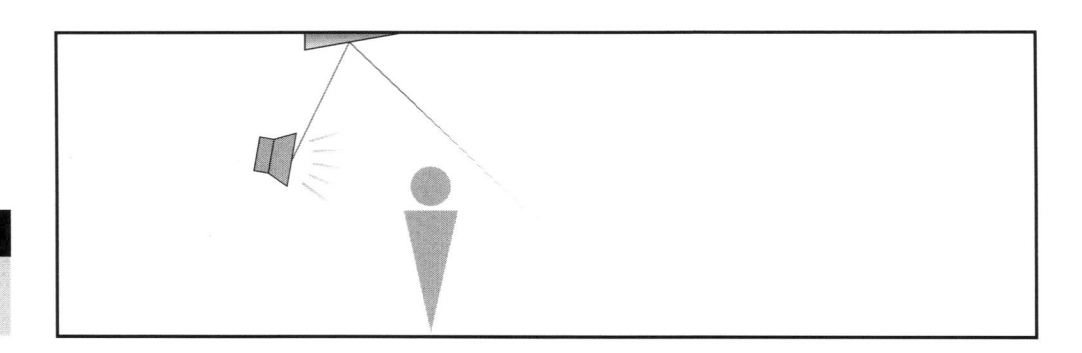

2.3.2.2 Bass Trapping

You can attenuate room-modal problems by using bass traps. These devices are used to smooth out the low-frequency response of the room, resulting in cleaner and tighter low end. Bass traps come in many different forms and shapes: two main systems used are pressure and velocity devices. In essence, pressure devices are placed against walls and attenuate low frequencies when they hit the boundary. These devices are hard to implement without the help of a professional acoustician, as they need to be built to specifically attenuate a particular problem frequency. Velocity devices, on the other hand, can attenuate a greater range of low frequencies; thus they are often called broadband low-frequency absorbers. Velocity devices are thick porous devices such as mineral fiber panels placed at a small distance from a wall. The larger the air gap between the panel and the wall, the lower is the frequency range being attenuated. A 10 cm to 15 cm air gap can help a lot in extending the low-end attenuation of bass traps. No matter what room size is being treated, there is no such thing as too much bass trapping, especially in smaller rooms, where most walls should be treated with bass traps.

Since room-modal problems are created between parallel walls, corners of the mixing room will exhibit more problems by building up bass from the two different dimensions of the room. As such, bass traps work best when placed in wall-to-wall corners, wall-to-ceiling corners, and less

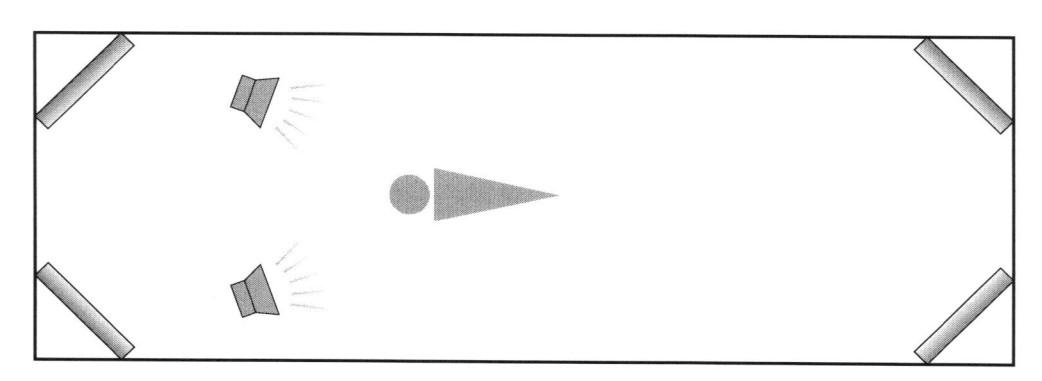

FIGURE 2.11
Bass-trap placement: top view.

FIGURE 2.12

Bass-trap placement: side view.

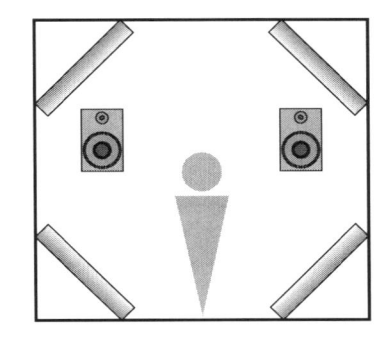

FIGURE 2.13

Bass-trap placement: back view.

commonly (because of the extra floor space needed) wall-to-floor corners (Figures 2.11–2.13). It is important to note that the common soft acoustic foam sold as corner bass traps does very little on its own to attenuate bass. When the foam is placed behind a more rigid panel in corners, it can extend the low-end absorption, but it affects only low-mid-frequencies if used on its own.

Symmetry is not important for bass trapping, as low-end frequencies are nondirectional. Because they are, if you have a limited supply of bass traps, a trick to find the best corners to place them in is to play low-pass-filtered (at about 300 Hz) white or pink noise and place a sound-pressure-level (SPL) meter in the different corners of the room. Using noise rather than pure tones ensures that all low-end frequencies are being represented for the SPL meter to "hear." The corner giving the highest reading should be treated in priority.

You can also do bass trapping on flat walls, although this remedy will only reduce the room-modal problems of the wall it is placed on. A good way of ensuring that the ceiling/floor modes are reduced as much as possible is to use thick ceiling clouds above the listening position, placed with an air gap above them (Box 2.1). Taking this step has the advantage of treating the ceiling for both bass and first reflections to the listening position. You could do the same on the side walls when you are treating first reflections.

AUDIO PRODUCTION PRINCIPLES

> **BOX 2.1**
>
> *A cheap way of creating an acoustic ceiling cloud is to adapt an old trick used to tighten up drums in a recording: suspending a large upside-down umbrella filled with acoustic foam over the listening position.*

2.3.2.3 Absorption versus Diffusion

The choice of using absorption or diffusion (Figures 2.14 and 2.15) when you are treating mid- and high frequencies depends on the room's reverb time to be achieved. Absorptive materials reduce the amplitude of the problem frequencies, while using diffusion scatters those frequencies across the room at lower amplitude.

Using a mix of these two types of treatments can adjust the room-size perception. This is very hard to judge without the use of room-analysis software displaying a waterfall view of the reverb time by frequency (Figure 2.16).

It is important to reduce the control room's standard reverb time (RT60) to just below half a second, while not allowing too dead a sound. Mixing in a room that has too much or too little reverb can be quite challenging. Overall, the RT60 should be even across the spectrum; waterfall graphs can be useful in figuring out whether absorption or diffusion should be used. Note that the smaller the room, the shorter the RT60 needed and the more absorption should be used. A good starting point is to cover roughly 50% of the room with absorption materials.

Since high frequencies are directional, the front wall behind the speaker does not need to be treated with anything other than bass traps. The back wall of the mixing room does, however, need to be treated. As mentioned previously, the type of treatment to be used there depends on the size of the room. In small rooms, absorptive materials are often preferred,

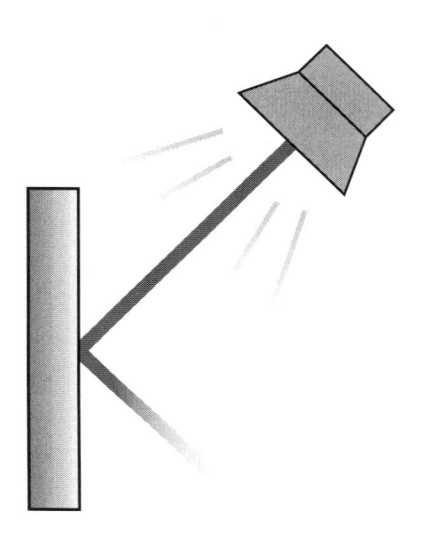

FIGURE 2.14
Absorption.

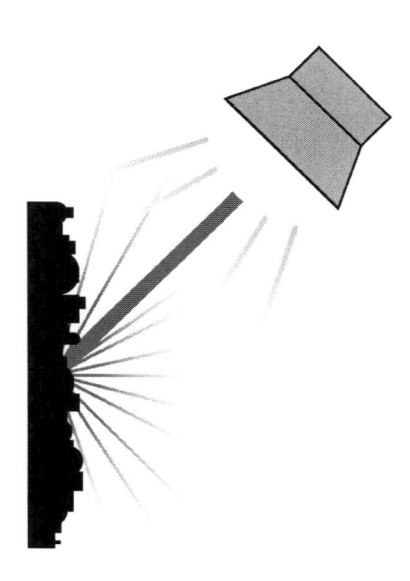

FIGURE 2.15
Diffusion.

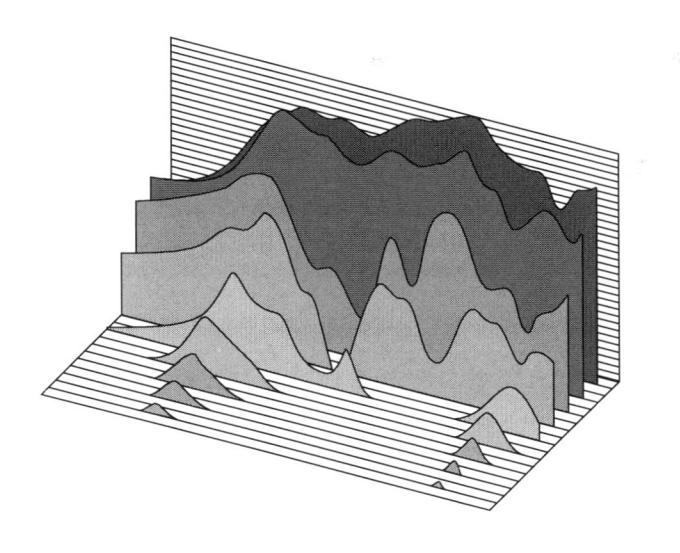

FIGURE 2.16
Reverberation time: waterfall view.

while in larger rooms reflective materials work best. In rooms where the listening position is at least 3 m from the back wall, using the "live end/dead end" (LEDE) technique can work well (Figure 2.17). This method involves using mostly absorption in the front half of the room, while using diffusion on the back wall.

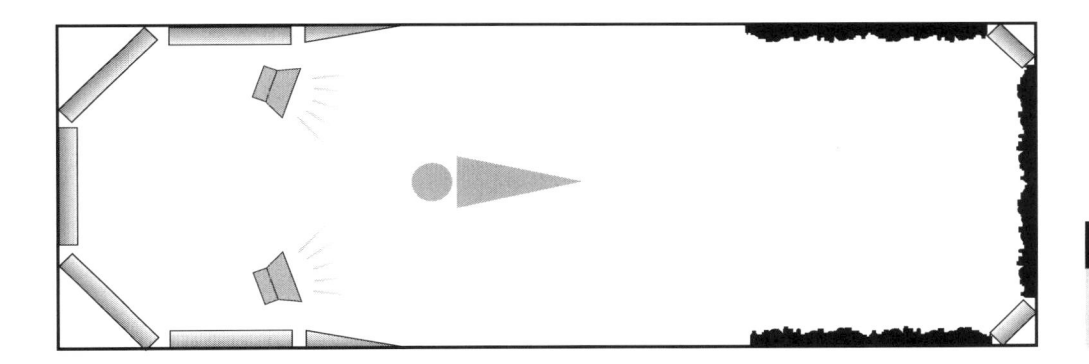

FIGURE 2.17
"Live end/dead end" method of acoustic treatment.

The exception is if you hear flutter echoes at the back of the room. In this case, and since flutter echoes are caused by high mids and mid-frequencies bouncing off parallel walls, you should use absorption on the side walls behind the listening position.

2.3.3 Room-Analysis Software

Good room-analysis software can be an invaluable tool in order to get the best result from the acoustic treatment you are using and to properly position the monitors. At the very least, the software should be able to display the frequency response of the monitoring system as well as a waterfall graph of the room's different reverb times per frequency. The aim when moving monitors and subwoofers, and placing the acoustic treatment is to achieve the flattest frequency response at the listening position. If you are using room-analysis software, a special test microphone is required to ensure that the recorded response is uncolored by the equipment. Test microphones can be quite cheap; the cheapest is the ones built into SPL meters. Note that SPL meter microphones often roll off in the low end and are therefore not suitable for judging room-modal issues (i.e., subwoofer placement and bass-trapping needs).

When you are doing measurements, it is important that the room be set up as it would normally be used during mixing. This precaution means keeping any doors and windows closed or open (depending on normal use), equipment racks placed at their normal position, and the microphone set at ear height, placed at the listening position, and facing the ceiling to ensure it is not angled toward one speaker or the other. A fairly long sine sweep (around three seconds) should be used to ensure that the room is "excited" in a similar way that it would under normal mixing conditions.

When you are reading the different graphs and measurements, it is important to use the least amount of smoothing on the response curve, to ensure that an accurate representation is displayed. For mid- and high frequency, you can use up to 1/12th-octave smoothing (Figure 2.18). For low-frequency readings and subwoofer placement, you should use 1/48th-octave smoothing (Figure 2.19). Note that while the aim is to achieve a flat response curve, it is normal to see a slow decline above 1 kHz or so. The amount of high-frequency roll-off will be dictated by the amount of absorptive materials used in the room.

FIGURE 2.18
Room-analysis graph: 1/12th octave smoothing.

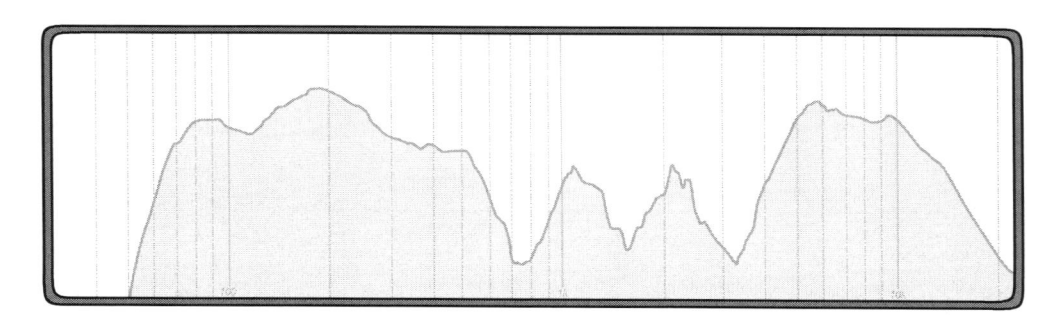

FIGURE 2.19
Room-analysis graph: 1/48th octave smoothing.

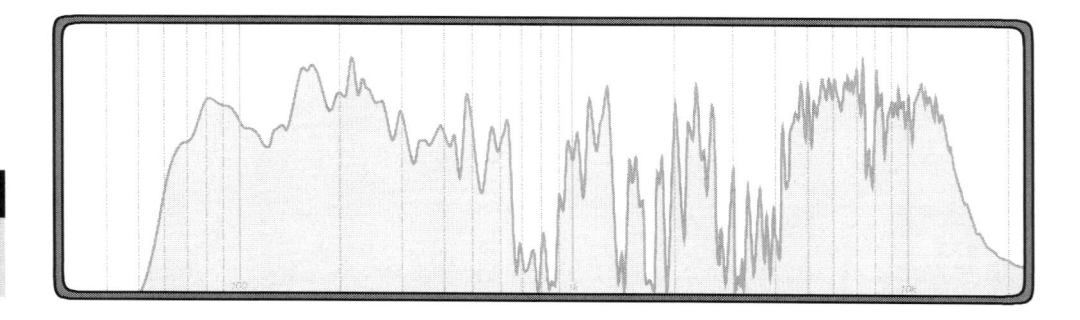

When using waterfall graphs, ensure that the full RT60 is displayed by allowing for 60 dB of decay over the length of the sweep (or longer) to be shown. Waterfall graphs can be very useful in figuring out which type of treatment to use: absorption or diffusion. The rule of thumb is to use absorption in the frequency areas that are slower to decay and diffusion in the frequency areas that have shorter decays. For example, if the low frequencies decay a lot more slowly than high frequencies, using more bass traps, less acoustic foam, and more diffusers is needed. On the other hand, if the high frequencies decay more slowly than lows (quite rare), angling walls to reduce room-modal problems and using more acoustic foam for high frequencies would work better.

If you are working on a budget without a room-analysis software, it may be possible to display a rough estimate of the room's frequency response by automating a very slow sine sweep (around 20 s from 20 Hz to 20 kHz) and recording this sweep by using a test microphone at the listening position. In your digital audio workstation (DAW), it will then be possible for you to spot at which frequencies the peaks and dips occur by looking at the automation value. While this way is not very accurate for analyzing a room, it will still show if there are big issues at the listening position (Figure 2.20).

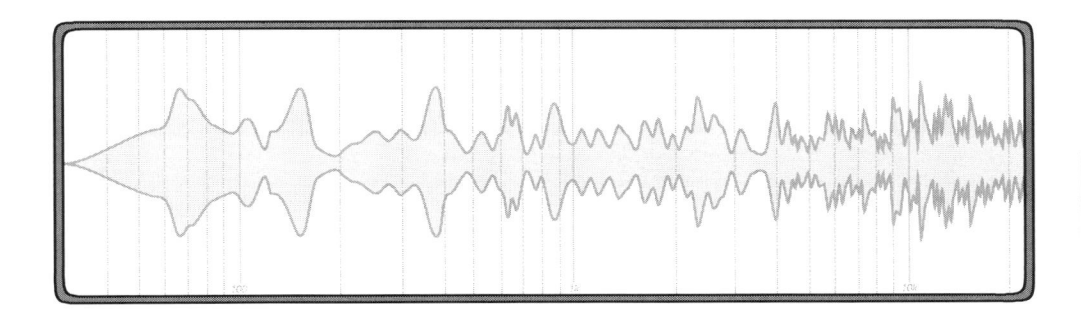

FIGURE 2.20
Recorded waveform from sine sweep (do-it-yourself technique).

Once you have done all of the analysis, correctly positioned the monitors, and used the necessary acoustic treatment, you should achieve around 10 dB of difference between peaks and dips. Less difference is great, although very hard to achieve without professional studio construction. More than 15 dB of difference is not acceptable in a mixing studio environment.

2.3.4 Room-Correction Software

Once you have optimized the whole monitoring system through speaker placement and acoustic treatment, you can further optimize the room by using room-correction software. These programs often come as plug-ins to be placed on master faders, software altering the output of the sound card, or hardware systems. Note that such software is intended to be only the "cherry on top" and cannot fix large problems. This sort of software uses reverse EQ curves from what it has analyzed in the room to flatten the overall response even more. Thus it does not alter the phase of the signal and is therefore ineffective at re-boosting frequency dips created by phase cancellation in the room. It is important to note that no more than about 6 dB of correction should be used with these systems.

Chapter 3

Preproduction

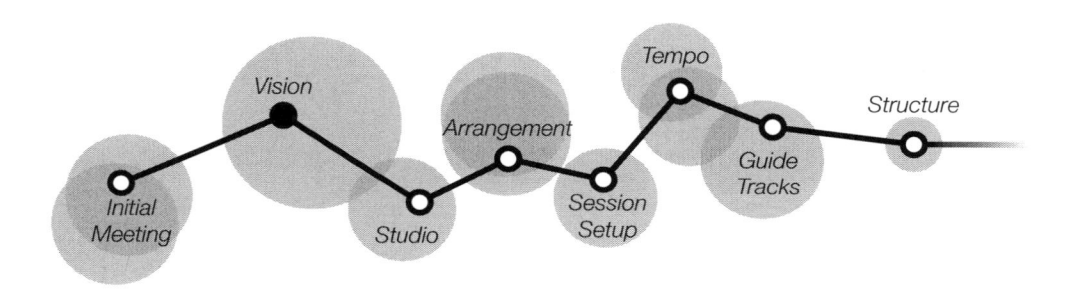

Preproduction may be the most important stage of a recording project, and yet it is too often neglected by young producers and engineers. Everything that can be done early on in the project should be dealt with straight away. Every minute spent in preproduction will save an equal or greater amount of time in the following stages of the project. Do not make the mistake of overlooking this step: plan, prepare, and deal with issues as they arise.

So what is involved in preproduction?

The first step is to meet with the musicians and attend rehearsal sessions. This is the time to decide which songs or pieces are to be recorded and which of them are potential singles. As producer, you will need to identify who is the "alpha" member of the band or ensemble (if there is no designated leader or conductor) and ensure that this person becomes your ally. Now is also the time to set a vision for the project. A clear vision will not only help you decide on technical aspects of the recording, but it will also help during tracking and mixing when creative choices are left for the producer or engineer to make. This may be the most crucial part of the project! For instance, if a band's drummer and guitar player think they are in a metal band, the bass player is playing funk, and the singer believes the band is a pop-rock outfit, pleasing everyone throughout the project is going to be a difficult task. You should ask other questions before checking which genre they see themselves fitting, as some musicians prefer not to categorize their own music. Ask about the places they play live; ask to see their social network pages, websites, and previous CD artwork, for example.

This research should give a clear idea of how the band or ensemble members see themselves and will help with creative decisions in the studio should you need to suggest that, for instance, a band feels more like a nu-metal outfit rather than pop rock. Or you may need to give an opinion if an a cappella chorus has decided on repertory but not on tempo, dynamics, or balance. Do not get in the studio before everyone has agreed on the vision! A lack of agreement will only create tensions and misunderstandings during the project.

Now is the time to ask the band all the relevant questions that will ultimately dictate the recording studios and techniques used, the mixing studios and techniques used, and even the mastering studios where the project will be finalized. When you are recording a smooth disco band, a studio with a Neve desk could be more appropriate than one with a Solid State Logic (SSL) desk when recording the drums. On the other hand, if the band members have agreed that they are more on the pop-funk side of things, an SSL-equipped studio may be just what the recording needs to give the required punch to those drums. A clear vision of where the project is headed should also dictate whether extra instruments will be required on the recording or not. A punk-rock outfit may not need to have a string quartet breakdown on their single, even if the drummer's sister can play the violin! Only if the band has decided on a vision will you be able to reject such a request if you feel that it would not suit the song. Editing also greatly varies, depending on the genre of music. While industrial-metal drums are usually hard quantized and sound replaced, jazz drumming is rarely modified in terms of timing and tone.

The goal of the editing/mixing engineer is to pick up where the recording engineer left off, and bring the project closer to a finished product that carries the vision of the artist. This transfer of labor means that the mixing engineer has lot of creative control over the song. For example, he or she is responsible for picking the right drum takes. Do you want a drum fill every four bars or only at the end of sections of the song? Should the second chorus sound the same as the first one for consistency's sake, or should it have more power in the guitars? Are we using two doubles of the vocals, or twenty? These are all creative decisions that will greatly affect the overall feel of the song. A pop-rock song can be turned into a heavy-metal one with only a few changes to the mix! All of these decisions will be dependent on the vision. Having the producer in the room during mixing could be a great help for making those kinds of decisions but is not always possible, practical, or necessary: for example, if a producer charges a lot of money for his or her mere presence or stops the mix engineer every ten seconds to ask about the process being done. A far better option would be for the mixing/editing engineer to speak with the producer and the band before the mixing session to ensure that the vision will be followed.

It is important for the producer to listen to any previously recorded material that the band may have. This research will give extra information on how tight the musicians are in the studio, as well as the sonic quality that the next recording will have to beat. No band gets in the studio to record a product that is of lesser quality than its last recording. It is important for the producer and engineer to know what sonic quality the band is expecting, and not to accept to work on the project if they do not believe they can produce a record that is of equal or greater quality than the previous recording.

Once everyone has a clear idea of how the project should sound, it is time to prepare for the recording phase. While the vision will help decide whether a live recording or overdubbing each instrument is more appropriate, now is the time to clear this decision with the band. Start thinking of such issues as where each instrument should be recorded, time frames, and budget. I do not list all the tasks that a producer is responsible for, since there are already many excellent books out there that cover this topic. Instead, let's look at preproduction from an engineer's point of view. To save time (and money), you should arrive at the first recording session with the following already prepared:

3.1 DAW Session Preparation

Set up an empty session with all tracks to be recorded named, routed, grouped, and color coded. For example, when you are recording a three-piece band (drums, bass, guitar), the session should contain all the drum tracks already set up and named, all routed into a drum bus, and all color coded. The same also applies to bass and guitars in that example. You should already know how many microphones will be used on each instrument and how many layers of each instrument will be recorded. While recording in layers rarely applies to drums and bass, it is not uncommon to record more than one guitar part to achieve the well-known "wall of sound" effect. If you are familiar with the studio, you should also write down which microphone and preamplifier combination you plan on using during the session. Separate headphone mixes should also be set up in advance to save time in the studio. Taking all of these steps before getting to the studio will make the recording session run a lot more smoothly (and cheaply!).

It is very important to build good habits when it comes to session file management and display settings. "Clean" and well-organized sessions often sound better. This approach obviously has nothing to do with the quality of the audio, but everything to do with the engineer's attention to detail. If you don't do the simple tasks of color coding, adding markers for sections of the song, proper labeling of tracks, and proper routing (Figure 3.1), how do you expect to spend the time required to finely adjust a kick-drum sound?

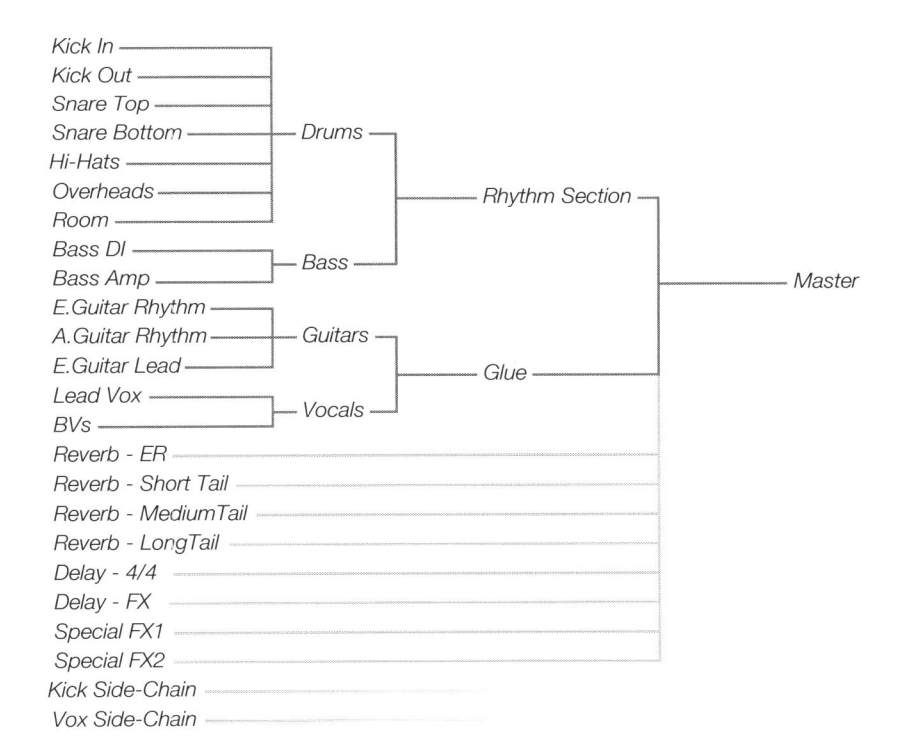

FIGURE 3.1

Organization of tracks within the digital audio workstation.

3.2 Tempo

While you are recording the guides for each song, you will need to set the session's tempo. Everyone must agree on the tempo from the start, as it is very hard to change it after the final recordings. One way to figure out the tempo is to play a click track through loudspeakers during a rehearsal and get everyone to agree on a specific tempo while playing to it. You may also want to take into consideration how the musicians are feeling on the day. If they are hyped about the recording, they may end up rehearsing at a faster tempo than usual. If they are tired, they may play more slowly. Another option is to record the guides and quantize them to the closest tempo value. Modern DAWs allow for the master tempo to be changed and the audio tracks to be time stretched with it. While the quality may not be great, the band can then decide on what sounds best before recording the final takes. Since this decision should be made early on, make sure to try different tempos and pick the most appropriate. The producer should have the final say about choosing the right tempo.

3.3 Structure

In much the same way that a session with blank tracks must be created, use the guide tracks to write down the different sections of the song (Figure 3.2). Run through this list with the band members, as they may have different names for sections of the song (bridge, middle eight, break, solo are all names commonly used for a C part in a song for example). Use the terms that the band members are comfortable with so that there is clear communication during the recording sessions.

FIGURE 3.2 Song structure.

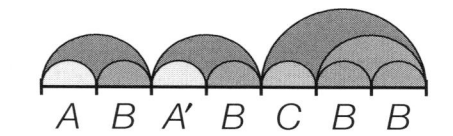

3.4 Guide Tracks

If the project is not to be recorded live, all guide tracks should be recorded beforehand. This setup usually means rhythm guitars, bass, keyboard, and vocals. All of these guides should be recorded to a click track unless the aim is to have a "tempo free" recording. Editing instruments will often take longer when the tracks are not recorded to a click.

With the exception of good drummers and musicians who regularly practice with a metronome, most people hate playing to a click. There are a few tricks to get a musician to give a tighter performance when he or she is playing to a click, such as ensuring that the headphone mix is just right, or sending the click to one ear only. Another trick is to set up a drum loop that somewhat resembles what the drummer will play throughout the song. The guides can then be recorded to this drum loop (Box 3.1). All the band members should be able to play in time as they stop focusing on an unfamiliar click sound.

> **BOX 3.1**
>
> *Pro Tools is excellent for setting up a session, as the built-in drum machine (Boom) already has preset rhythmic patterns that will play when a single key from C3 to D#4 is pressed. Some other DAWs have similar features that allow for a quick setup of a drum pattern, but if yours does not have this feature built in, make sure to have a range of loops ready to be dragged in at any moment to keep the session flowing.*

3.5 Sound "Map"

With the vision set for the project, a "map" that lists all the different sounds present in the song should be created. This map can take the form of a series of adjectives describing each individual sound as well as the full mix. Noting down what each instrument should sound like may uncover the fact that not all instruments will be able to have a "bright and punchy" tone, for example, or that there are too many bass instruments in the project. This list should then be expanded to the most suitable equipment to be used to achieve the envisioned sounds. Included in this map should be an instrument list mentioning make, model, and different options for the given instrument. For example, if the electric guitar sound is to be clean, twangy, and a little hollow and thin, it might be decided to use a Fender Stratocaster with light-gauge strings and its pickup set to the out-of-phase position between the bridge and middle pickups and run through a Vox AC30 amplifier. When you are faced with an instrument or sound that you have little knowledge of, research can help in narrowing down the list of different models available. The rest can be achieved during the recording session. A similar approach is often helpful when you are using synthesizers during sound design. A clear view of how the frequency content of the sound evolves over time can help in deciding the type of oscillator, filter, modulation, and effects to be used.

3.6 Studio Hierarchy

It is important to differentiate the roles that engineers and producers are responsible for in a project. While the line separating the two is often blurred on projects in which the engineer acts as producer (and vice versa), in a session in which there are clearly two different persons, each with a set role assigned, the line should be a little more rigid to avoid the "too many cooks in the kitchen" issue. Here are some of the roles typically assigned to the different personnel involved in the studio:

- Assistant engineer: Preparing the session, setting up microphones, doing line checks, patching, and following direct explicit requests from the head engineer

- Head engineer: Deciding on the best methods and equipment for achieving a particular sound, interpreting requests from the producer
- Producer: Realizing the vision of the project through any and all means necessary. This role often involves looking at song structure and arrangement and refining lyrics, all while referencing other songs in similar genres.

A simple example of this division of labor could be could be the following:

PRODUCER: We need a raw, piercing, and distorted electric guitar sound.
HEAD ENGINEER: We will use equipment X and Y with technique Z on the amp.
ASSISTANT ENGINEER: I'll set up equipment X and Y with technique Z, and ensure that everything is working.

It is also important to remember who "works" for whom, and ensure that this hierarchy is not threatened by ego or overenthusiastic studio members. For a general rule, the assistant engineer works for the head engineer, the head engineer works for the producer, and the producer works for the band (more specifically, the "alpha" member of the band). In order to ensure that the project is as smooth as possible, studio members should discuss project-related matters only with their direct "supervisor." For example, having the engineer suggest modifying a part while the band is in the studio will only create tension, as the producer will be left not only to manage the band's reaction to this request, but also the engineer's. In this case, waiting for a break and having a one-on-one discussion between engineer and producer can make this suggestion a lot less stressful. Another example would be the assistant engineer suggesting a different tone to be achieved altogether or a different way to achieve the required tone. Once again, this kind of suggestion is a sure way of creating tension in the studio, as it (often innocently) bypasses the next person in the "line of command."

3.7 Budget

Another role of the producer is to create a budget for the project and ensure that it is followed correctly. While the budget should be set prior to the preproduction process, it is important to mention it here. Drafting two budgets for the project—a low-cost option and a full-production option—is generally preferable. The reason for this approach is that it gives the band the broadest spectrum of options, which can then be discussed to see the areas where extra money should be spent in order to get the best possible outcome.

A typical budget will include the following items:

- Producer fee: It should be broken down per hour for smaller projects, per song for projects with undefined scope, or per project for those set in stone. Do not forget to plan for preproduction time!
- Studios/Engineers: It should include studios and time used for recording each individual instrument, song mixing, and mastering. If a particular engineer is to be used at any stage, his or her fee also needs to be displayed.
- Session musicians/equipment hire: If the project needs a particular musician or equipment, list how long for and how much this addition will cost.
- Relocation costs: In some cases, relocating to a different city for the project may be beneficial (for credibility, networking opportunities, motivation, creativity, access to specialized engineers). If this is the case, all associated costs such as travel, accommodation, and food should be listed.
- Consumables: Depending on the nature of the project, some consumables may need to be itemized. For example, when you are recording to tape, rolls need to be included. If the band wants a hard drive with all of the session data at the end of the project, this cost needs to be included.
- 10% padding: Always include this padding on top of the final budget to cater for unforeseen circumstances. More often than not, this extra will be needed, and if not, the band will love you for it!

***Chapter* 4**

Microphone Basics

Microphones come in all shapes and sizes. Since there are a lot of books that cover microphone types and techniques in great detail, I am only sharing a few tips that can help speed up microphone choice.

4.1 Choosing the Right Microphone

When choosing a microphone, keep the following in mind:

4.1.1 Frequency Response

It is important to understand how the frequency response of the microphone reacts to the frequency content of the instrument being recorded. For example, it is not always good to match the most "open sounding" microphone with a bright sound source, as this choice may result in a harsh recording. It is often preferable to match a dull microphone with a bright source or a bright microphone with a dull source.

4.1.2 Transient Response

The same guideline as in "Frequency Response" applies. It is often best to match a microphone that rounds off transients with instruments that are transient heavy.

4.1.3 Polar Pattern/Directionality

This factor is critical when you are recording more than one instrument at a time, or recording in a very reverberant room. Different polar patterns also exhibit different tonal qualities in terms of the frequency content you want to capture.

4.1.4 Physical Size

While this is not directly related to the recorded sound, the physical size of the microphone may affect the performer. For example, is the microphone size and position going to be restricting?

4.1.5 Cost and Durability

The most expensive and fragile microphones are often left in the recording studio, whereas the more durable ones are used in live sound reinforcement.

In a recording studio situation, the frequency response, transient response, and polar pattern are the main aspects that will dictate the appropriate microphone to be used.

4.2 Microphone Types and Attributes

First off, it is important to note that different microphone designs carry general "sonic signatures." The three main designs used in recording studios are dynamic, condenser, and ribbon microphones.

4.2.1 Dynamic Microphones

- Rounding off of transients
- Less detailed in the top end
- Robust
- Most are cardioid
- Cheaper
- Can handle high sound-pressure level (SPL)

4.2.2 Condenser Microphones

- Detailed top end
- Require phantom power
- Fragile
- Sharp transient response
- Multiple polar pattern
- May require a pad for high SPL

4.2.3 Ribbon Microphones

- Smooth top-end roll-off
- Sharp transient response
- Extremely fragile, especially when exposed to bursts of air
- All are bidirectional
- Can handle high SPL
- Some require phantom power, while others break when phantom is applied

4.3 Polar Patterns

Polar patterns can also play an important role in the final sound you are recording. The three main polar patterns are cardioid, bidirectional (or figure eight), and omnidirectional.

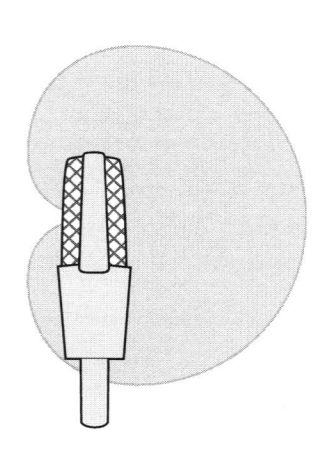

4.3.1 Cardioid

- Subject to proximity effect (which sometimes relates to a "warmer sound" than that of other polar patterns)
- Least subject to room sound when close microphone technique used
- Most subject to differences in tone when recording off-axis

These attributes and directionality are illustrated in Figure 4.1 and Audio 4.1.

FIGURE 4.1
Cardioid polar pattern.

4.3.2 Bidirectional

- Subject to proximity effect
- Great rejection of sound from the sides
- Open sound, although some models tend to emphasize mid-range frequencies

These attributes and directionality are illustrated in Figure 4.2 and Audio 4.2.

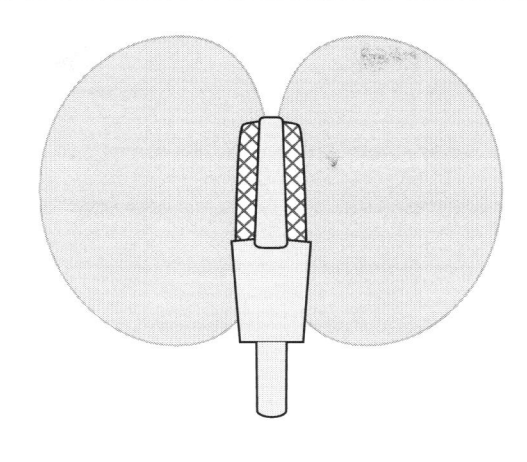

FIGURE 4.2 Bidirectional polar pattern.

4.3.3 Omnidirectional

- Natural and open sound (extended top end)
- Subject to bleed, room tone, and natural reverberation
- Extended low-frequency response, although not subject to proximity effect

These attributes and directionality are illustrated in Figure 4.3 and Audio 4.3.

FIGURE 4.3 Omnidirectional polar pattern.

4.4 Physical Size

Different microphone sizes exhibit different tonal qualities. While dynamic and condenser microphones have fairly consistent sounds when you are comparing small and large diaphragms, the tone of ribbon microphones, although often said to be "vintage", changes with models and is therefore not mentioned here.

4.4.1 Small-Diaphragm Dynamic Microphones

These microphones usually have a pronounced midrange response and tend to round off transients. Their frequency response extends to low mids and some bass, but rarely will they pick up sub-frequencies. The same applies to the other end of the spectrum: they can pick up treble, but they may not be the best choice if you are looking for "air." Microphones such as the SM57 are well known for picking up the "meat" of midrange instruments such as guitars and snare drums.

4.4.2 Large-Diaphragm Dynamic Microphones

These microphones are commonly used for bass instruments; their frequency response extends to sub-bass frequencies (see Box 4.1 for additional uses). Different models can have radically different tones. For example, a Shure Beta

52 can sound great on a kick drum, as the mic has a natural presence boost to bring out the detail of the beater hitting the skin, but could potentially sound thin on a bass cab for the very same reason.

> **BOX 4.1**
>
> *While most engineers use large-diaphragm dynamic microphones for low-end instruments, they can be used for midrange heavy sounds such as brass or even vocals.*

4.4.3 Small-Diaphragm Condenser Microphones

Not including some very special designs, these microphones are typically used to pick up anything but sub-bass frequencies. They can be thought of as more detailed versions of dynamic microphones (with a top end extending to the upper limit of our hearing). Depending on the design, these mics can sometimes be more directional than their large-diaphragm counterparts.

4.4.4 Large-Diaphragm Condenser Microphones

These microphones often have the widest frequency response and tend to be less directional than their small-diaphragm counterparts. These mics are the design of choice for recording vocals.

4.5 Distance

The distance between the microphone and the sound source can greatly affect the recording. There are a few variables to consider when you are choosing the right distance for a given situation. These factors will affect frequency content as well as transient information in the sound.

4.5.1 Frequency Content

First off, and related to the polar pattern selected, proximity effect can be used to generate more low end in the recording. Up to roughly 20 cm, a buildup of bass will occur when you are using cardioid or bidirectional patterns. Another factor that can affect the perceived low end in the recording is the decrease of high frequencies with increased distance, as air slightly affects the phase of high frequencies. Note that humans naturally increase the amount of high frequency in their voices with distance in order to use less power and therefore project better.

4.5.2 Transient Information

As distance increases, transients and waveforms tend to be rounded off (Figure 4.4). This effect is partly due to the previously mentioned loss of high frequencies as well as the comb filtering happening from sound waves hitting room walls and blending with the direct sound. In much the same way, reverberation increases with distance and gives clues to the listener with regard to the recording environment.

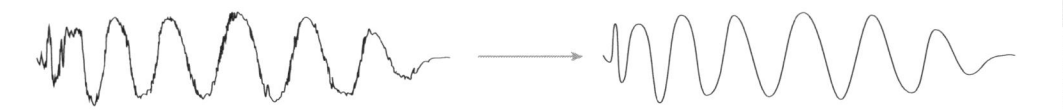

FIGURE 4.4 Effect of distance on waveform.

Overall, it can be useful to use the distance in the recording as it is meant to sound in the mix. If an instrument is going to end up at the back of a mix, record it from a distance. If on the contrary, it is meant to be at the forefront of the mix, place the microphone close to the sound source.

4.6 Orientation

The result of facing a microphone directly toward the sound source sounds different from that of having the microphone "off axis." This result is true even for omnidirectional polar patterns. It is helpful to think of the orientation of the microphone as a tone control, similar to the tone knob of an electric guitar.

4.6.1 On Axis

Sounds tend to be more powerful when recorded on axis (Figure 4.5; Audio 4.4 ⏵). Finer detail can be captured, as most of the direct sound is being recorded.

FIGURE 4.5 On-axis recording.

4.6.2 Off Axis

Angling microphones off axis usually results in a duller sound (Figure 4.6; Audio 4.5 ⏵). Since high frequencies are very directional, the "detail" of the sound being captured runs past the microphone capsule without being fully picked up. The frequency response of a microphone also changes, depending on its angle to the sound source. This response does not mean that instruments should always be recorded on axis, though. Dullness can be used to your advantage when you are trying to make instruments fit within a busy arrangement.

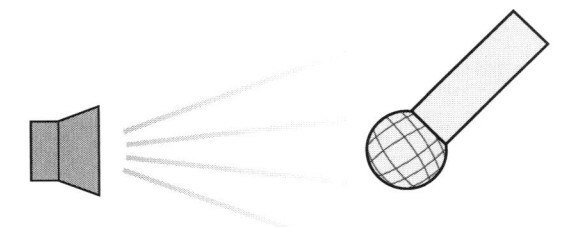

FIGURE 4.6 Off-axis recording.

4.7 Other Considerations

While not directly related to microphones, the shape of the recording room, construction materials of the walls, and placement of the sound source within the room all affect the final recorded product. The shape and construction materials of the room mainly affect the reverb quality and therefore the transient information of the sound source. As an example, a small room full of absorptive materials will sound rather dead compared to a large room covered with reflective surfaces. The placement of the source within the room mainly affects the frequency content of the sound source. Examples include placing the sound source in a corner to increase low-frequency content or, on the contrary, placing the sound source away from room boundaries to generate a more natural sound.

4.8 Microphone Selection Recap

Selecting a microphone for a given sound source should be primarily done by using frequency and transient response. It is helpful to get in the habit of matching opposite microphone to sound-source quality. For example, you should match an instrument with a lot of transients and high frequencies, such as a heavily strummed steel-string acoustic guitar, with a microphone that smoothes out transients and rolls off high frequencies (through design or off-axis positioning). Another example would be that of a dull voice. In this case, matching it with a microphone that has a fast transient response and emphasized high frequencies can bring life to the recording.

Note that the polar pattern you use also plays a significant role in the sound you achieve. When you are selecting a polar pattern, the two main attributes to take into consideration are frequency response (through the use of proximity effect or omnidirectional bass boost), and ambience to be recorded.

Table 4.1 shows the attributes of different microphone types, polar patterns, and placements.

TABLE 4.1 Microphone Types, Polar Pattern, and Placement: Sonic Attributes.

	Frequency Response (Emphasis)			Transient Response		Recorded Ambience	
	Low	Mid	High	Fast	Slow	Open	Closed
Dynamic (Card.) Close	•	•			•		•
Dynamic (Card.) Far		•			•	•	
Condenser (Card.) Close	•	•	•	•			•
Condenser (Card.) Far		•	•	•		•	
Condenser (Omni.) Close	•	•	•	•			•
Condenser (Omni.) Far	•	•	•	•		•	
Condenser (Fig. 8) Close	•	•	•	•			•
Condenser (Fig. 8) Far		•	•	•		•	
Ribbon (Fig. 8) Close	•	•		•			•
Ribbon (Fig. 8) Far		•		•		•	
All Microphones On Axis	•	•	•	•			
All Microphones Off Axis	•	•			•		

4.9 Stereo Recording Techniques

While there are dozens of different stereo recording techniques, only a few are used to record popular music instruments such as drums or guitars. Each technique has advantages and disadvantages that you should take into consideration when deciding on microphone placement.

4.9.1 Spaced Pair (AB)

This technique works primarily by introducing time differences between microphones to achieve the stereo spread (Figure 4.7; Audio 4.6 ⓟ). It is very effective at capturing instruments that are physically large in relation to their distance from the microphones. For example, a marimba can be big if the microphones are placed close to it, or small if the microphones are placed a few meters away. This technique works especially well when you are recording instruments that contain mainly low mids and low frequencies. The overall sound resulting from this technique can be described as "spacious" and "lush," but may lack a sense of focus and adequate localization.

Because of the time differences between each microphone, some phase displacements will happen. While this effect is not an issue for instruments that are placed on the ground and do no move, it can introduce uncontrollable phasing when you are recording instruments that move during performance, thanks, for instance, to an overenthusiastic acoustic guitar player (Audio 4.7 ▶). This displacement results because a moving instrument will change the phase relationship between each microphone over time. Generally speaking, the three-to-one rule, which states that the microphones should be three times as far from each other than they are from the sound source, should be followed to minimize phase issues when you are using this technique. Note that the sweet spot of a recording made with a spaced pair of microphones is smaller than that of other techniques due to the listener needing to be halfway between the speakers for a true representation of the instrument to be displayed.

The technique is rather ineffective when you are dealing with instruments that are small in relation to their position from the microphones. For example, a drum kit consisting of only kick, snare, and hi-hats or an acoustic guitar recorded from a few meters away will effectively arrive at both microphones at the same time. This placement will therefore reduce the time differences and result in a somewhat mono recording.

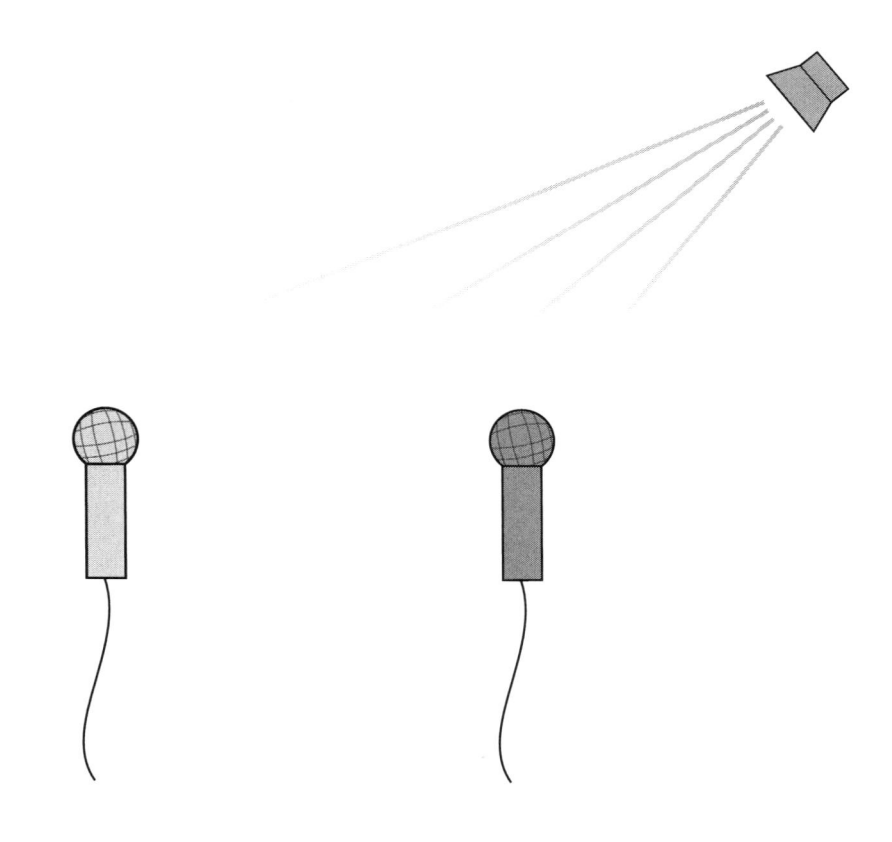

FIGURE 4.7
Time differences to achieve stereo.

4.9.2 Coincident Pair (XY)

This technique works primarily by introducing amplitude differences (both general and spectral amplitude) between microphones to achieve the stereo spread (Figures 4.8 and 4.9; Audio 4.8 ▶). Using a pair of cardioid microphones angled at 90 degrees, this placement is very effective at capturing instruments that are physically small. Angling the microphones away from each other can increase the perceived width of the instrument. This technique works especially well when you are recording instruments that contain mainly high mids and high frequencies. The overall sound of this technique can be described as "clear" and "focused," with good localization of instrument.

Since this technique uses coincident microphones, it introduces very little time differences and phasing issues. Thus it is excellent for mono compatibility, as well as recording instruments that move during performance (Audio 4.9 ▶).

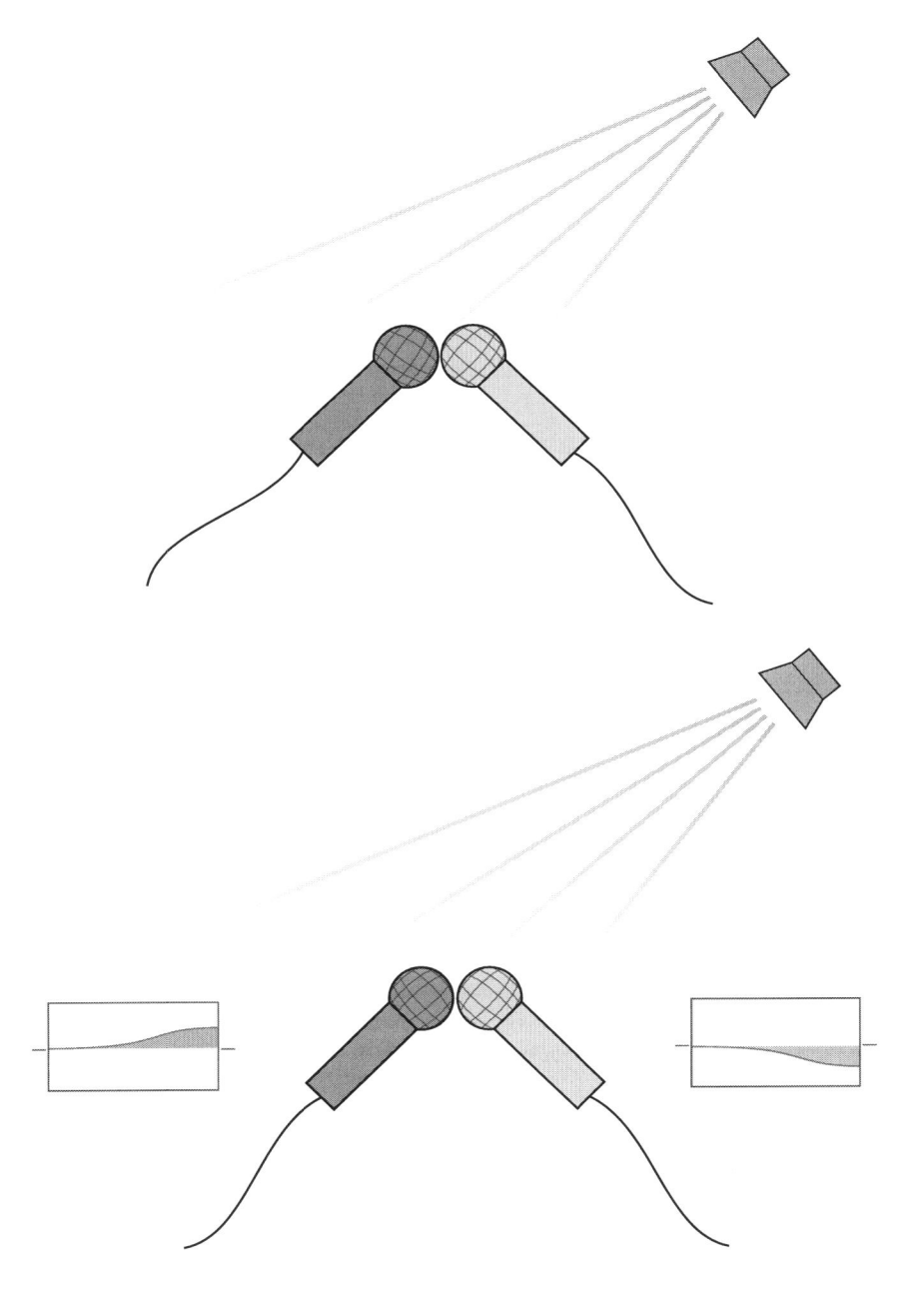

FIGURE 4.8
Amplitude differences to achieve stereo.

FIGURE 4.9
Spectral differences to achieve stereo.

4.9.3 Coincident Pair (Blumlein)

This technique is very similar to XY, since it also uses coincident microphones (which must be angled at 90 degrees from each other), but set to a bidirectional polar pattern. All of the benefits of an XY pair are also applicable to this technique, with the added advantage of being able to more accurately capture physically larger instruments (Audio 4.10 ▶). Since the microphones also record information from their back, recording in a good sounding room is very important when you are using this technique.

4.9.4 Coincident Pair (Mid-Side)

This technique also uses coincident microphones, which makes it fully mono compatible. Just like an XY pair, it is very good at recording instruments that move during performance, as well as instruments that are physically small (Audio 4.11 ▶).

This technique has the added advantage of being able to modify the stereo width post-recording and often ends up sounding wider than an XY pair. Setting up a mid-side (MS) pair requires a center microphone (different polar patterns give different sounds and stereo spread), and a bidirectional side microphone. The center microphone must be panned in the middle, while the side signal should be duplicated and panned hard left and right, with one side polarity reversed. The center microphone can be the following:

4.9.4.1 *Cardioid*

This technique emulates the sound of an XY pair by using hyper-cardioid microphones (Figure 4.10; Audio 4.12 ▶). It has good back rejection, is subject to proximity effect, and has an off axis coloration that seamlessly transitions from the cardioid mid mic to the bidirectional side mic. This center microphone should be used when you need to achieve a more focused sound. When you are mixing, if the mid microphone is louder than the sides, it has more off axis coloration, and the emulated XY pair's angle reduces. If the side microphone is louder than the mid microphone, it will have more on axis coloration, and the emulated XY pair's angle increases.

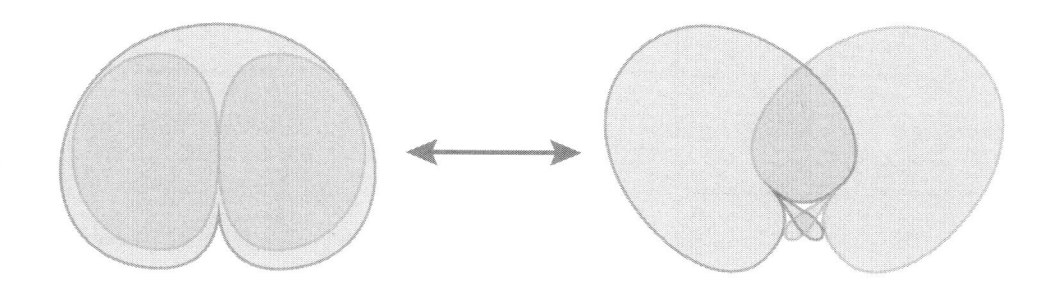

FIGURE 4.10
MS pair emulation with cardioid center microphone.

4.9.4.2 Omnidirectional

This technique emulates the sound of an XY pair by using cardioid microphones angled at 180 degrees, facing away from each other (Figure 4.11; Audio 4.13 ▶). It picks up more room ambience than using a cardioid center microphone, is not subject to proximity FX, and displays only slight off axis coloration given by the side microphone. This center microphone should be used when the instruments are placed on either side of the bidirectional microphone. When you are mixing, if the mid microphone is louder than the side, it reduces the on axis coloration, and the emulated XY microphones' polar patterns become more omnidirectional. If the sides microphone is louder than the mid microphone, it will increase on axis coloration, and the emulated XY microphones' polar patterns become more hyper-cardioid (i.e., more directional).

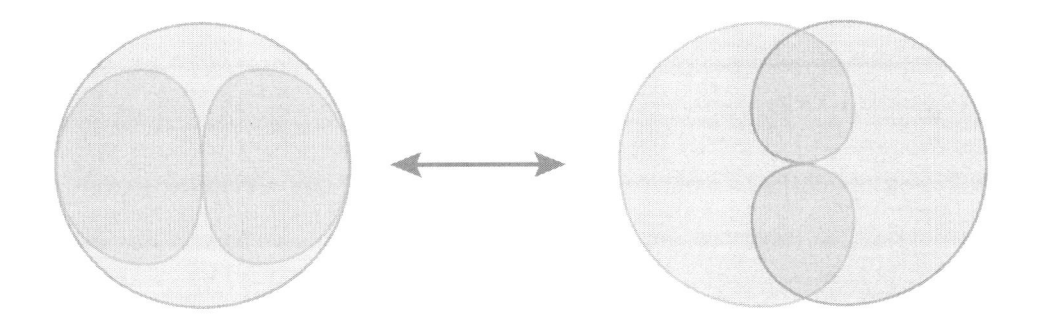

FIGURE 4.11
MS pair emulation with omnidirectional center microphone.

4.9.4.3 Bidirectional

This technique emulates the sound of a Blumlein pair of microphones (Figure 4.12; Audio 4.14 ▶). It provides greater stereo spread and a more even reverb response, and is subject to proximity FX. During mixing, if the mid microphone is louder than the side, it increases off axis coloration, and the emulated Blumlein pair's angle reduces. If the side microphone is louder than the mid microphone, it will increase on axis coloration, and the emulated Blumlein pair's angle increases.

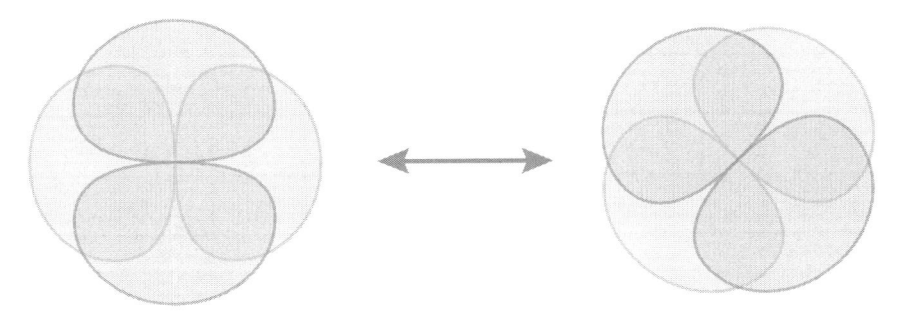

FIGURE 4.12
MS pair emulation with bidirectional center microphone.

4.9.5 Near-Coincident Pairs (ORTF, DIN, RAI, NOS)

The different near-coincident pairs available such as ORTF (for Office de Radiodiffusion Télévision Française), DIN (for Deutsches Institut für

Normung), RAI (for Radiotelevisione Italiana), or NOS (for Nederlandse Omroep Stichting) aim at emulating how the human hearing system works (Audio 4.15 ▶). They use time, amplitude, and spectral differences to achieve the stereo spread and are often used as good "in-between" solutions when none of the previously discussed techniques work. The different methods mentioned all use cardioid microphones spaced between 17 and 30 cm apart, and facing away from each other at an angle between 90 and 110 degrees.

If you cannot get a decent recording from an unusual instrument, place it where you believe it will sound best in the room. Once you have done so, walk around it and try to find the place where it sounds best to your ears. Then place a near-coincident pair of microphones where your ears are. All you need to do now is to listen to those microphones in the control room and get the assistant to change their angle and spread until the best sound is achieved from that position. Since near-coincident pairs of microphones are supposed to emulate the human head and ears, this technique works well for most acoustic instruments when other techniques have failed to give good results.

4.10 Stereo Recording Technique and Polar Pattern Selection Recap

With the amount of information given in this chapter, it may be necessary to review the different variables presented to an engineer and deduct the most appropriate stereo recording technique for a given situation. The options, in order of importance, are as follows:

4.10.1 Room Quality and Ambience to Be Recorded

How does the recording room sound? If you are recording in a good-sounding room, the choice of an "open" or "closed" recording is available. If you are recording in a bad-sounding room, only a "closed" recording can be achieved, as you should aim to record as little room sound as possible.

> Good/Open
> Bidirectional or omnidirectional should be the preferred polar patterns, and the microphones can be placed far away from the sound source.
>
> Bad/Closed
> Cardioid should be the preferred polar pattern, and the microphones should be placed close to the sound source.

4.10.2 Quality/Localization to Be Achieved in the Recording

What is the intended focus to be achieved in the recording? Techniques that use amplitude differences to achieve stereo tend to sound "strong" and "focused," while techniques that use time differences to achieve stereo tend to sound more "lush" and "sweet."

> Focus
> XY, Blumlein, or MS should be the preferred techniques.

> Lush
> Spaced pair or ORTF should be the preferred techniques.
> Note: Blumlein and MS (with bidirectional or omnidirectional center microphone) techniques can also give a lush sound to recordings, as they also record the time differences coming from the back of the room.

4.10.3 Size of the Instrument

When you are thinking about the size of the instrument relative to the position of the microphones, is the instrument small or large? If you imagine that your eyes are positioned where the microphones will be placed, how big does the instrument look? For example, when you are looking at an acoustic guitar from 10 cm away, it will be large. Look at the same guitar from 2 m away, and it will then be small. Another way of thinking of this variable is to consider whether the three-to-one rule mentioned earlier can be followed. If it can, then the instrument is large. If it cannot, the instrument is considered small.

> Small
> XY or Blumlein should be the preferred techniques.

> Large
> Spaced pair or MS should be the preferred techniques.
> Note: ORTF could also work as a third choice for both small and large instruments.

4.10.4 Movement of the Instrument during Performance

Is the instrument moving during the performance or is it static? For example, a drum kit is static, since it rests on the ground, while an enthusiastic brass section used to playing for a live audience is likely to be moving.

> Static
> All stereo recording techniques can be used.

> Moving
> XY, Blumlein, or MS should be the preferred techniques.

4.10.5 Frequency Content to Be Achieved in the Recording

Is the aim of the recording to sound full with extended low frequencies, or somewhat thinner with a flatter low-end frequency response? This variable directly relates to being able to use proximity effect in cardioid and bidirectional microphones, or using the low-end emphasis of omnidirectional microphones if recording from further away.

Full
Omnidirectional should be the preferred polar pattern for recording far away from the sound source, and cardioid or bidirectional should be the preferred polar patterns for recording close enough to the sound source to utilize proximity effect.

Thin
Cardioid or bidirectional should be the preferred polar patterns for recording far from the sound source.

See Table 4.2 for a summary of options.

TABLE 4.2 Stereo Recording Techniques and Polar Pattern Selection Matrix

	Room Good Bad [G – B]	Quality Focus Lush [F – L]	Size Small Large [S – L]	Movement Static Moving [S – M]	Frequency Full Thin [F – T]
AB (Omni.) Far	•		•	•	•
AB (Omni.) Close	•		•	•	•
AB (Fig. 8) Far	•		•	•	•
AB (Fig. 8) Close	•		•	•	•
AB (Card.) Far	•		•	•	•
AB (Card.) Close	•		•	•	•
ORTF Far	•		• •	•	•
ORTF Close	•		• •	•	•
Blumlein Far	•	•	• •	• •	•
Blumlein Close	•	•	• •	• •	•
XY Far	•	•	•	• •	•
XY Close	•	•	•	• •	•
MS (Omni.) Far	•	•	• •	• • •	
MS (Omni.) Close	•	•	• •	• • •	•
MS (Fig. 8.) Far	•	•	• •	• • •	•
MS (Fig. 8) Close	•	•	• •	• • •	•
MS (Card.) Far	•	•		• • •	•
MS (Card.) Close	•	•		• • • •	

4.11 Creating Interest with Microphones

There are many ways to create interest with microphones, and while there are no real rules here, the key is to introduce movement in the tone being recorded. By making sounds evolve or change throughout different parts of a song, you can create interest in the mix.

One very easy way to create interest is to use different microphone combinations, positions, and polar patterns for different parts of a song. For example, when recording a guitar cab, you could set up three microphones, but use only two at any given time during the song. You can do the same when recording a snare drum by recording the whole song with two microphones (same model on axis and off axis, or different models) but only ever using one at a time during mix-down. This approach will give you a chance to subtly change the tone of the snare during different parts of the song. You could even fade one in and the other out during the course of a verse, for example, for a slightly evolving snare sound (the tracks must be phase aligned for this technique to work). For vocals, try recording the verses with a cardioid polar pattern, and choruses with an omnidirectional pattern. This technique will have the effect of making the verse vocals sound warm and close to the listener, and the chorus to be more "open."

The position of microphones can also add a lot of movement and be used for special effects. For example, adding a microphone in the airlock of the studio can give another option during a drum fill or pre-chorus, for example. When it comes to position, anything could work; it is a matter of listening and trying out ideas. Note that "unusual" microphones can also work wonders in creating special effects throughout the recording process. Examples of these effects include rewiring a telephone's earpiece and using it as a microphone for the "original telephone FX," or patching the listen-mic output from the console to the DAW to record a heavily compressed sound from the recording room.

One final and necessary note for this chapter is that it is sometimes helpful to use microphones as virtual EQs: that is, use complementary microphones on a sound source and blend their tone to achieve the desired tone. A classic example is recording vocals with two microphones: a large-diaphragm condenser, and a ribbon microphone. At the mixing stage, blending the warm tone of the ribbon with the presence of the condenser instead of simply equalizing one microphone can work very well in obtaining the right tone.

Chapter **5**

Recording

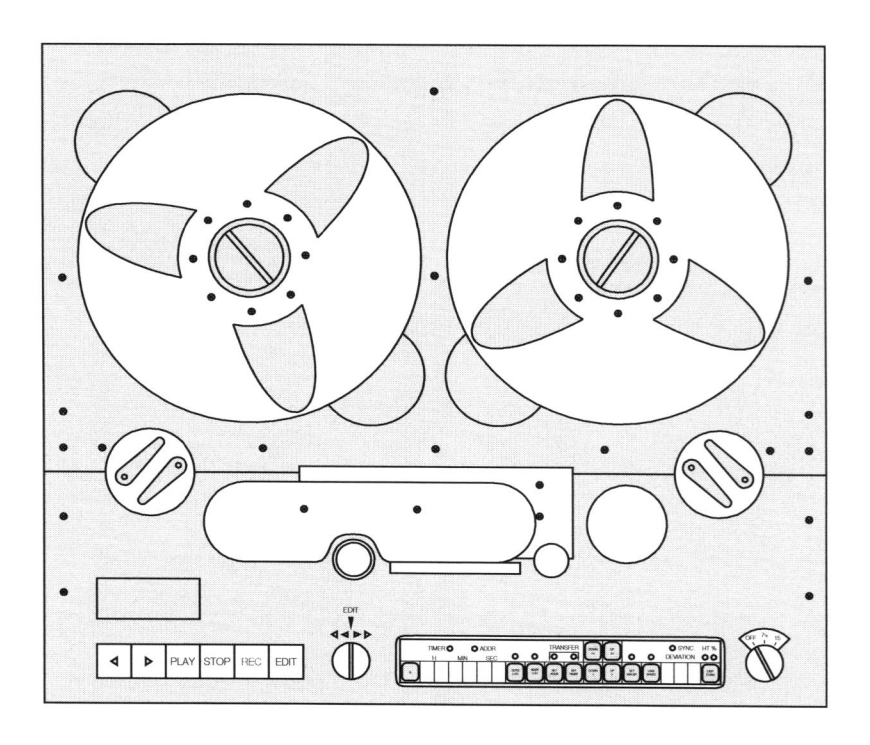

Much can be said about audio engineers and their lust for high-end equipment. It is important to remember that equalizers, compressors, and other audio processing units are just tools to make the task of audio engineering an easier and more enjoyable one. Great recordings were made long before DAWs and fancy digital reverb units, just as number-one hits are made with laptops and no outboard equipment each year. Differentiating what tools you need as opposed to those that you want is crucial in making the right choices when you are setting up a working environment.

5.1 Recording Medium

Most engineers record in DAWs these days. While some studios still use tape, it is more for the tone that it imparts on sounds than by necessity. All DAWs are similar in the way they can record, play back, and mix audio. Some have specific features that make them more compatible with certain genres of music, but a song can be mixed to a professional level with any of the major DAWs available.

5.2 Gain Staging

It is important to observe proper gain staging when you are recording to ensure that the correct tone is recorded. In a rule of thumb, 24-bit (and 32-bit float) recordings should be done at around −18 dBfs. This setting ensures that the equipment being used is not driven unnecessarily. If the studio has been properly calibrated, this level should utilize the full dynamic range of analog equipment, as well as give headroom for random loud sounds to come through undistorted. If you are working at 16 bit, you should record the sounds to be as loud as possible without clipping the analog to digital converters (ADCs). For better understanding of gain staging, it is important to understand the different settings available and common issues arising from digital recording.

5.3 Bit Depth

Since bit depth is directly related to dynamic range, it is important to understand what the different values available have to offer. In essence, 16- and 24-bit processing work similarly, as they are both fixed-point processing. The only difference is that the higher value offers a greater dynamic range (approximately 1 bit = 6 dB). Sixteen bit offers a theoretical dynamic range of 96 dB and 24 bit offers a theoretical dynamic range of 144 dB. In practice, 16 bit offers 84 dB of usable dynamic range and 24 bit offers 132 dB of usable dynamic range because the two least significant bits are reserved for the sign and noise floor. If you consider that the best ADCs available have a dynamic range of roughly 130 dB (with 120 dB of dynamic range being a more realistic usable figure when including the dynamic range of other equipment used for recording), when you are recording at

24 bit, −12 dBfs should be the highest level at which to record sound. In fact, using one less bit (or recording at −18 dBfs) is considered something of a standard. The reasoning behind this value is twofold. The first point regarding the dynamic range of analog equipment has already been made above, but note that microphone preamps have nonlinear characteristics in the way they color sounds. More specifically, their nominal operating level (when the volume-unit [VU] meter reads "0") is not the highest level at which they can operate. It is commonly accepted that going a couple of decibels above 0 VU can sometime add character to a sound, but exceeding this value too much starts adding unpleasant distortion. For this reason, and to mimic the "analog way" of recording and mixing audio, you should consider −18 dBfs as the nominal operating level of the DAW and recording/mixing around this value to be necessary (Figure 5.1). Nowadays, many plug-in manufacturers emulate this system by introducing harmonic distortion in the sounds run through them above −18 dBfs.

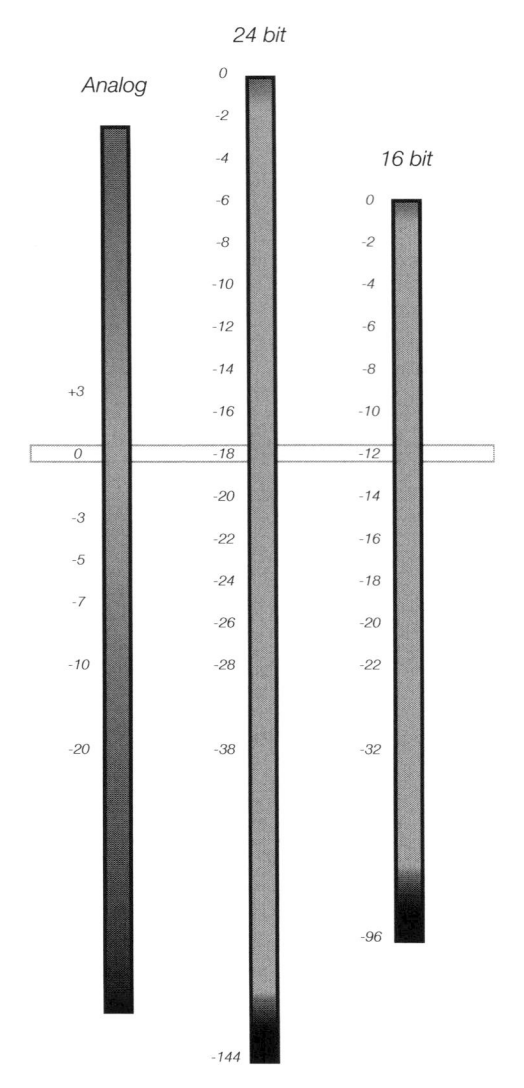

FIGURE 5.1
Operating levels in the analog and digital domains.

Floating-point processing such as 32-bit float is a somewhat different beast. In a 32-bit float file, 24 bits are fixed (23 bits for the mantissa and 1 bit for the sign), and 8 bits are floating (called the exponent). The true benefit of using floating-point processing is the safety that it brings. Using 32-bit float means that no digital clipping done inside the DAW is permanent. Even if a channel is clipping on output, for example, turning down the master will fix the introduced distortion. Another example would be to export a 32-bit-float audio file with the master fader pushed to clip, and reimport it in a different session. The volume of the file can be brought down to restore the original unclipped audio. Note that while it is nearly impossible to clip the track itself, plug-ins that emulate hardware units often also emulate the sound of the hardware unit when it is pushed above its nominal operating level. These plug-ins often start distorting at −18 dBfs, and this saturation cannot be undone when you are turning the next device down in volume.

Another benefit of using floating-point processing is that there is no rounding of values when you are using digital processors; thus the highest quality is kept at all times. While recording audio does not benefit from being done higher than 24 bit (fixed), given that ADCs work at 24 bit, once recorded, processing should be done at 32-bit floating point. This processing will allow for better-quality audio in the final product. Some DAWs such as Pro Tools 11 (and subsequent versions) automatically process the audio at 32-bit float, even if the original file is 24 bit (although elastic audio in the case of Pro Tools requires the file to be changed to 32-bit float manually before processing for the best quality to be achieved).

5.4 Sample Rate

Modern equipment offer a range of sample rates available for recording, processing, and playing back audio. The commonly used sample rates are 44.1 kHz, 48 kHz, 88.2 kHz, 96 kHz, 174.4 kHz, and 192 kHz. While there has been a lot of research regarding the pros and cons of using high sample rates in a recording, it is commonly accepted that the higher the quality of the equipment being used, the less difference there is between the different sample rates. This relation means that a very well-designed ADC running at 48 kHz can sound just as good as the same converter running at 96 kHz. The difference for manufacturers is that it is harder to implement a good-quality anti-aliasing low-pass filter for the converters at the lower sample rates, as it will need to be a lot steeper.

Note that the quality of audio recording and playback is very much affected by the quality of the system's clock. The amount and type of jitter in the system is just as important as the quality of the converters!

5.5 Aliasing

Put simply, aliasing is frequency content that is unrelated to the original audio signal added below the Nyquist frequency (half of the sampling-rate frequency). Aliasing can happen at the different stages of production and care should be taken to avoid it as much as possible. During the recording stage, aliasing can happen because the filter used in the ADC is not working properly, or the recorded signal is too loud. In any case, the harmonic content created above the hearing range could be "folded back" onto the hearing range and can create enharmonic distortion.

In order to avoid aliasing, you must observe proper gain staging. As mentioned in Chapter 11, "Synthesis Basics," square waveforms contain odd harmonics, and since clipping will square off waveforms, extra higher harmonics are created. These higher harmonics, even above 20 kHz, will then be folded back onto the hearing range.

The other way to avoid aliasing is to use higher sampling rates when you are recording and mixing. Since aliasing is frequency content added below the Nyquist frequency, using higher sampling rates and therefore moving the Nyquist frequency higher mean that the aliasing will be added above the human hearing range and therefore become a nonissue. To increase the sampling rate during recording, you must increase the DAW's whole session sampling rate. During mixing, you can use oversampling in plug-ins to achieve the same effect by working at higher sample rates (and therefore avoiding aliasing within the plug-in), and then bringing the sampling rate back down to the session's sample rate.

5.6 Recording Live versus Overdubbing

Before booking a studio to record, you must decide whether to record the band live or overdub each part one by one (or a combination of both). Both methods have pros and cons, but the decision to go with one or the other largely depends on the genre of music and preference of musicians being recorded, and should have been decided during preproduction.

Recording live is often the preferred method when the genre of music requires a natural and organic feel to be conveyed. In the studio, just as in live performances, musicians feed off one another's energy when they are performing. This interaction is what is often called "feel" or "vibe" and is an essential part of a recording. Overall, the groove tends to be better in genres such as funk, reggae, and jazz when everyone is recording at the same time. Musicians can "lock in" to one another's grooves a lot more easily when they can read body language rather than relying on only what they hear coming from headphones to give them a sense of groove. In this case, there may not be a need for a click track (once again, depending on the genre); not using a click track will solve many issues if the drummer is not tight enough as the band deals with tempo drifts together while playing. If the recording project is planned to be done live, musicians will have a tendency to rehearse more, as they understand that "fixing mistakes in the mix" is not an option. This attitude can be a double-edged sword, though: in a live recording situation, the overall quality project is dictated by the "weakest link." Everyone must be well rested and "in the zone" for a good live recording to happen. If one person is sick, the whole band will feel and sound sick. If someone in the band is not good enough, there is no way of replacing that person with a session musician (or a virtual instrument in the case of drummers and bass players). While these issues will be very clear during preproduction, some bands still decide to go ahead with a recording project while fully knowing that one player will bring the whole project down. If one player makes a mistake, everyone needs to rerecord the take. This necessity can kill the vibe in minutes and ruin a full day of recording. The musicians must be a lot tighter and well rehearsed when recording live.

However, if time in the studio is restricted, recording live can be the best option. From a budgetary point of view, less time spent going back and forth to the studio for overdubs often means cheaper projects. Depending on the scale of the project, however, this approach could end up being more expensive, as recording at the biggest and best studio in town for a few days may cost more than overdubbing in appropriately sized studios. Small studios often have the same excellent microphones, preamps, and outboard processors as the big studios. For the price of one day in a big studio, a two- or three-day deal could be done in small ones. For recording overdubs, less equipment is often needed, and the same sonic quality can therefore be achieved in smaller studios. Do you want to save even more money? Get the bass and guitar players to record their parts in their own time by using a high-quality direct-input (DI) box and ADC. They will be able to try, edit, and bring the parts to a studio that has the amp/mic/preamp/outboard FX necessary, and re-amp them all in one go. The same concept applies to the equipment needed for the project, as the technical requirements are usually greater in a live recording session: a bigger recording space and more equipment such as microphones, preamps, compressors, and EQs are needed. If the equipment available is limited, concessions will need to made. In the case of overdubs, each instrument can be recorded in a different space, and the best possible equipment can be used for the task at hand. A studio might have a great drum room or a very dead vocal booth, for example. While overdubbing in separate rooms means that achieving a sense of togetherness will be a little harder, it helps a great deal in separating instruments. More exotic recording spaces such as corridors and stairways are also available when you are working this way. The drums could even be recorded in a studio that has a great live room and an SSL mixing desk; the bass can be recorded at a band member's house by using a DI. The next step could be to move to the other side of town for a studio that has the biggest collection of guitar amps available, and finally have the vocals done at the local mastering studio for its clean signal path and high-end processors. Note that not being able to use the best microphone, preamp, EQ, and compressor on every single instrument can be a good thing. Instruments will be easier to separate when the recorded sounds do not carry the same sonic characteristics.

If complete control over the recorded sounds is necessary, overdubs are generally the best option. Some genres of music such as pop or metal require heavy post-processing from audio engineers. For example, because of the bleed introduced when you are recording live, making edits and tonal adjustments to one instrument will also affect others. Recording a singer and an acoustic guitar at the same time means that brightening one instrument will almost always mean brightening the other. If the singer already has a harsh voice or the guitar is piercing to begin with, this brightening can become a big issue. Of course, this technique can be used in the opposite manner. For example, one EQ can be used to adjust the tone of multiple instruments at once! Separating instruments for post-processing may be a little more difficult because of bleed. Although the use of EQs, gates, baffles, distance, rejection side of microphones, and so on can greatly reduce spill, it will be nearly impossible to completely avoid in a live situation. Overdubs allow for every part to be as tight as possible. Not only can you rerecord parts a number of times until you achieve a great take, but you have the added advantage of being able to edit previous parts before you record the following ones. Overdubbing helps with tightening the production as a whole, as every new part gets recorded to an already spot-on performance. Different parts can be recorded and compiled for the best possible outcome. For example, the first drum take might have a great verse, while the pre-chorus drum fill in take #2 has more power than the others. In much the same way, small guitar licks can be cut from different takes and put together to create an optimal performance. Recording with bleed does have one major advantage, though: a sense of togetherness can easily be achieved. The instruments in the mix gel more easily when everything has been recorded at once. When the elements of the mix are individually compressed and equalized, instruments affect one another because of the bleed being processed, therefore creating synchronized movement between them without using bus compression (sometimes called glue compression).

From a psychological perspective, both recording types have their advantages. When recording in overdubs, musicians can record on days that best suit them. Recording on days that are inconvenient for band members (recording on a Sunday evening when a band member works early on Monday, for example) will never yield the best results, as the members get stressed. Overdubs mean that all the players can select a time when they know they will be in the right headspace. There is generally less pressure on musicians recording this way, because they have the ability to rerecord if anything goes wrong and achieve the best possible outcome. New ideas can also be tried in the studio. Even though specific parts such as drum fills or special FX should have all been agreed on during preproduction, it is not uncommon for musicians to try different parts when there is no need for the whole band to play the part each time an idea is suggested. Once again, this flexibility could be a disadvantage, as it is then a lot easier for the bass player's girlfriend's cousin to make suggestions from the couch at the back of the room. More time could be wasted if the band tries every idea! The endless possibilities in terms of tones, parts, and instrumentation can make the recording project a nightmare, especially when there is no producer involved to mediate ideas. When you are recording live, there will be very few silly requests such as "can we add tablas in the second verse?" or "can we quadruple track that bass to make it thicker?" While experimentation is great when composing music, all of the extra parts should have been thought of during preproduction. Musicians can very easily hear when a live jam gets overcrowded with instruments and players, but that skill quickly disappears when it comes to recorded music.

There are a few advantages that are available only to projects recorded in overdubs. Recalling instrument settings to recreate the sonic characteristics of a recording is particularly feasible with only one instrument in the studio. It is a lot easier to remember the placement of a guitar cab within the room and microphone placement when it is the only instrument being recorded. Remembering the placement of each player within a room, baffles, amp and guitar settings (which are bound to be tweaked mid performance anyway!), and everything else will prove to be impossible in the event that the same sonic characteristics are sought during a later recording. Similarly, the engineer has the opportunity to perform "on the fly" automation with amp settings, and EQ settings, for example, and can thus give more life to the recording before the mixing stage. It is also very easy to get different tones from each instrument in different parts. For example, the drummer might use brushes for the first half of the verse, move on to side stick for the second half, and then use a different snare drum for choruses. Another example could be using one guitar amp for verses and another for choruses. These types of changes can prove very difficult at best when you are recording everything live. Communication with musicians can also be easier, as other band members are not fussing with their instruments while the engineer or producer converses with the person recording. On the other hand, during live recording, suggestions made to the band as a whole, even if aimed at a particular member, can be accepted a lot more easily than when directed straight at the musician.

5.7 Recording Order

While there are no rules when it comes to recording, when you are doing overdubs, here is the commonly accepted standard recording order for a "rock type" band:

- Guides
- Drums (and drum editing post-recording)
- Bass (and bass editing post-recording)
- Rhythm guitars (rhythm-guitar editing should be done during recording session to save time)
- Lead vocals (and lead-vocal editing post-recording)
- Backing vocals (and backing-vocal editing post-recording)
- Lead guitar (and lead-guitar editing post-recording)
- Fills, highlights, and sweetener (and editing)

Why?

Drums and bass are the rhythmic foundation of the song. They dictate most of the groove and dynamics of the song and should be recorded first so the rest can be recorded on solid foundations. The rhythm guitars are also an essential part of the song and finish the musical backdrop of the song. The rhythm guitars tend to be recorded after drums and bass, as they have a little more space to drift in and out of time (in a controlled way) and give more life to the recording. These instruments may play lead parts at some point in the song, but their main role is mostly supportive. After you have recorded the foundation of the song, it is time to record the most important instrument of all: the voice. There are very few genres of music that do not treat the voice as the main instrument. We always listen to vocals and lyrics more than anything else when we hear a song. Since this is the most important instrument, it should not be recorded after any other lead part. Once you have the vocal tracks, the exact recording order of the remaining instruments matters less. Since lead guitars are less important than vocals, recording them after will ensure that they are working around the vocals instead of the opposite. If vocals are recorded at the end of the project, the singer or singers will not be able to give their best performance, as the amount of space left for them in the arrangement has been reduced.

5.8 Mix While Recording

Start mixing instruments with outboard equipment as they are recorded. Having a clear vision for the project means that the intended sounds should already be clear, underlining why preproduction is so important. Why wait until the mixing stage, when another round trip of D/A and A/D conversion will be needed, to use outboard equipment? If there are high-pass-filter switches on your microphones, use them where needed. If the snare needs to sound large in the low mids, equalize it in during the recording session. You can make an exception if you are unsure of your ability as a mix engineer. If you do not know how to use a compressor, recording through one can cause more harm than good and cannot be undone!

5.9 Running the Session

There are a few general rules to follow when you are working with musicians in the studio. The first and most obvious is to ensure that the musician should be treated with the utmost importance. Since the aim is to capture the best possible performance, the engineer and producer should do everything possible to make the performer feel good. This goal often means giving positive feedback even when there is not much good in the performance.

Another important point is to build a rapport with the artist. While this relationship should have started at the preproduction stage, connecting with musicians on some sort of level can help with the smooth running of the session. Being able to talk the lingo of a particular musician has a lot of benefits. Knowing a few facts about the instrument being recorded is a great way of building trust with musicians. For example, being able to talk about the sound that different-size kick drums have or how a particular guitar model was used by a notorious musician shows the artist that you are interested and knowledgeable about his or her craft and helps a lot in building trust. If the engineer does not know much about the instrument being recorded, a good fallback strategy is to talk about studio equipment and how it will enhance the tone of the instrument. This approach effectively involves the musician in the recording process and helps building rapport during the session.

It also means ensuring that the session runs smoothly in terms of the technical aspects. There is nothing worse than having to delay a first drum take (for which the drummer was really pumped up and in the "zone") because one overhead microphone is not working. Everything should work perfectly when the musicians are in! There is much to be said about capturing the artist's performance at the right time. As musicians get prepared to record in the studio, there is always energy building up to a point where they will deliver their best performance. That best-performance window can sometimes be as small as a couple of minutes, so it is essential that the technical side of the recording does not prevent the session from running at the perfect time. It is the job of the producer and engineer to sense when the musicians are ready and hit Record at the right time!

A good habit to get into is to call the musician in the control room to listen to the recording after the first take. This step can further help build trust and ensure that the artist does not doubt the engineer's competency during the recording. After this first check, the artist will focus solely on his or her performance rather than think about the technical aspects of the session and whether the sound is being properly captured.

Another rule is to ensure that the musician currently being recorded does not feel singled out when he or she is performing. It can be demoralizing to see band members laughing out loud in the control room, even if the source of the laughter is completely unrelated to the recording. Always make sure that each musician feels comfortable. Sometimes it means dimming the lights in the control room so the performer thinks he or she is alone in the studio. Sometimes it means dimming the lights in the recording room so he or she knows that no one can see through. Sometimes it even means removing the band from the control room for a while.

To ensure that the performer does not feel neglected when he or she is in the recording room, never let more than thirty seconds pass without talking through the talkback system (obviously not when recording a take). Even if you have nothing to say to the performer, a simple "hang in there; we're just adjusting the gain on the preamp" or "we'll do another take in just a sec; I want to try something on your previous recording" can make him or her feel a lot more involved in the studio life and development of the song. Of course, the musicians do not need to know about every little adjustment being made on preamps or EQs, but if it has been more than thirty seconds since you last talked to them, hit the talkback button and say the first thing that goes through your head! Doing so shows that you care by including them in the studio conversations. It can work wonders in making them feel good about the recording process.

When working with a digital audio workstation, save different versions of the song. It is good to get in the habit of saving a new version of the session every time a major chunk of work has been done. For example, titles could be:

SongTitle_01_Guides
SongTitle_02_Drums
SongTitle_03_DrumEdits
SongTitle_04_Bass
...
SongTitle_12_MixPrep
SongTitle_13_Mix1
SongTitle_14_Mix2
SongTitle_15_BandRevisions
SongTitle_16_MasterPrep
SongTitle_16_RadioEdit

Working this way helps with retrieving previous takes, different guitar solos, drum fills, or particular sounds achieved during a specific mix session. A simple recall of the relevant session can help in retrieving parts to use in the latest version of the mix. Note that the last two numbers above are the same: the final mix should then be split into "full song" and "radio edit," as these labels refer to two different edits of the same mix.

5.10 Headphone Mix

The artist's headphone mix is a very important step in getting the best possible performance. Simply sending a split of the main mix to the headphones does not give enough flexibility to be able to pull the best sound possible. When you are setting a performer's headphone mix, it is important to monitor it as well so that the engineer is fully aware of what the performer is hearing when issues arise in the recording. Moreover, using the same model of headphones as what the performer is wearing ensures that the engineer is 100% aware of what is being sent to the live room. For full control over both the main and headphone mixes, ensuring that the sends are pre-fader allows the performer to retain his or her mix in the cans while the monitoring room can solo individual instruments as necessary.

You can use some tricks when setting up a headphone mix to steer the performance toward a particular goal. Examples include the following:

- Raising the volume of the headphones in parts of the song to push the artist to perform with more energy
- Raising the volume of the click in a drummer's headphones to help with timing
- Raising or lowering the volume of the singer's headphones to force him or her to sing harder or softer (or sing slightly sharp or flat)
- Raising the volume of the vocals in a singer's headphones to ensure he or she can hear him or herself and adjust the pitch correctly (taking one side of the headphones off can often achieve the same effect, too)
- Using similar FX on the vocals as what will be in the final mix to ensure that a singer projects his or her voice and performs in an appropriate way for the song

5.11 Pulling the Right Sound

There is a particular order that you should always follow when recording an instrument or designing a sound with synthesizers. When deciding on an order such as instrument/amp/mic/EQ/converters, always start at the source and work your way through to the ADCs. For example, when setting up a guitar for recording, decide on the tone in this order:

- Room geometry/construction materials
- Guitar model
- Guitar strings/plectrum type
- Guitar settings
- Amplifier model
- Amp settings
- Placement of the amp within the room
- Microphone choice
- Microphone placement and angle
- Outboard equipment (preamp, EQ, compressor)
- ADC

There is no point in tweaking an EQ before moving microphones around. There is no point in moving microphones if the tone settings on the guitar or amp are not the right ones.

The same applies in synthesis and sound design. Work your way from the oscillators to the final FX plug-in in the chain. Here again, there is no point in messing with the filters settings if the waveform is not the right one to begin with, just as there is no point in putting a chorus FX at the end of the chain until you have tried to achieve that effect by using detuned unison voices.

In the next few chapters, I give tips on recording the different instruments in a basic rock band. When faced with a more unusual instrument to record, try to understand how it generates sound and compare it to drums, guitars, bass, or voice. Ask yourself questions such as "Is this instrument transient heavy? Physically large? What is its purpose in the mix?" You may find that the techniques given in this book can work well for the situation.

Chapter 6

Drum Kit

Drum recording is often the biggest part of any project, since it usually requires the most equipment, space, and time. In this chapter, we look at the different requirements of drum recording, as well as some tricks to help run sessions smoothly.

6.1 The Room

When you are recording acoustic instruments, the room in which they are placed is a very important aspect of the final sound. The venue is particularly true with drum kits, for which capturing the overall sound is just as important as each individual drum.

Do not think that large rooms always sound bigger than small ones. Small rooms can often give the illusion of a larger kit because the many reflections from the walls increase the perceived sustain of each drum hit, giving something of a natural compression effect. The floor, ceiling, and wall materials also greatly affect the overall tone of the kit. Concrete surfaces will give a harder sound, especially as the midrange and high frequencies get reflected. Wooden surfaces tend to keep the initial attack of the sound but quickly damp the sound, giving a lush tail to every hit. Carpeted surfaces (depending on the thickness of the carpet and the underlying material) can choke sounds and absorb too much of the drum hits, resulting in a small-sounding kit. There are situations, however, that require a very dry and tight kit sound. In larger rooms, the surface materials are also important in reducing flutter echoes. In a rule of thumb, the softer the drum playing, the larger and more reflective the room can be used to capture "air" without sacrificing definition.

6.2 The Instrument

While tuning electronic and sound-replaced drums can be done quite easily during mixing, an acoustic kit must be tuned before the recording can start. For each song, the best scenario would be to tune the kick drum to the root note of the song's key, the snare to the root note or the perfect fifth, and toms to other consonant notes in the key used. This tuning will ensure that the rhythmic bed of the song works well with the harmony and melody of the other elements. While this sort of tuning is rarely done because of time constraints in the studio, spending extra time optimizing the tone of the kit before putting microphones up can help a lot in getting a solid final kit sound. Even if the individual drums are not in tune with the key of the song, ensuring that each drum sounds good on its own can go a long way.

A quick and dirty way of tuning drums is to dampen the bottom skin, hit the top skin, and tighten it until it sounds "in tune" with the song. Because the tension should be even across the skin, it is therefore necessary to lightly tap the skin next to each lug and adjust the tension until all areas of the drum sound the same. Damping the skin currently being tightened with a finger can help you discern the pitch more easily while tightening lugs. The next step is to tighten the bottom skin by using the same technique until it is at the same pitch as the top skin. For toms, keeping the same tuning for both top and bottom skins will result in a steady note that sustains well when the drums are hit. Tuning the bottom skin slightly lower will result in a pitch drop after each hit, keeping the same sustaining quality. Tuning the bottom skin slightly higher will result in a pitch drop after each hit, with less sustain (Audio 6.1 ▶). Also note that the way the toms are attached to the kit (using rubber brackets, metal poles, and so on) will affect their tone. It is therefore important to try different positions when the drums are set up.

For snare drums, this method can be somewhat modified to achieve a wider range of tones. Looser tuning can yield deeper tones (at the expense of a tight attack) compared to tighter tuning.

While this method can work well for snare and toms, it is often difficult to hear the pitch of a kick drum well enough to judge whether or not it is in tune with the song. In this case, tuning the front and back skins so they are just tight enough not to show any wrinkles should be enough to record one kick hit, load it into a sampler, and pitch-shift it an octave up to hear the actual pitch given. From there adjustments can be made on the kick drum, with constant checking of the results in the sampler can help achieve the right tone. Since the kick drum often gives the low-end foundation of the mix, it is important to ensure that the pitch is solid and therefore in tune with the song. In general, the correct tuning for kick drums should not be far from the starting position (around one or two turn of the tuning key).

There are many types of construction materials for drum shells (such as maple, birch, and fiberglass) which all have their distinctive sound. While it is not necessary to know how each type of wood or synthetic materials sound, if you have the choice between different drum kits, be sure to audition them to get the most appropriate sound for the song. Similarly, shell thickness can affect the tone of the drum, with thinner shells imparting more of their sound (as the shell vibrates more freely) than thicker ones. Thick shells vibrate less and are therefore more subject to the sound given by the drum head. The type of drum heads used will also affect the tone of the kit. In a rule of thumb, clear heads are a little brighter and ring out more when compared to coated heads, which sound a little warmer and tighter (Audio 6.2 ▶). The thickness of the drum heads can also affect tone. Thinner batter heads provide better attack than thicker ones, at the expense of thickness of sound and durability. For resonant heads, thicker ones provide longer sustain if they are not over-tightened.

The different sticks used will also change the sound quite dramatically. Nylon tips are a little brighter than wooden tips. Hot rods have less attack than sticks, a factor that can be great for removing hi-hat and cymbal harshness, but can also sound a little faint on snare and toms. Brushes and mallets are even softer than hot rods and both have their own distinctive sound (Figure 6.1; Audio 6.3 ▶).

FIGURE 6.1

Drum sticks, hot rods, mallets, brushes.

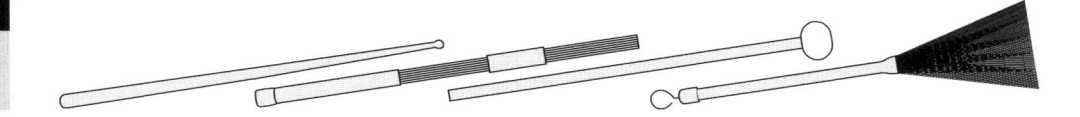

There are three different types of kick-drum pedal beaters: felt, wood, and plastic. Felt beaters have less attack than the other two. Both wood and plastic give very bright and punchy results, each with a slight difference in tone (Audio 6.4 ▶).

The snare drum often needs to be muffled so it does not ring out too much. While a large decay is great for a live snare drum, it is not usually sought after during recording. There are many ways to dampen a snare drum, such as using a doughnut ring, a very fine cloth on the snare, or tape. Note that damping the snare (and toms) will do very little in fixing the sound of an out of tune drum. The damping will merely try to mask tuning issues, similar to how reverb can sometimes hide a vocalist's pitch problems. The tone of the snare drum can be further modified by adjusting the tightness of the snares under the drum. Tighter snares give a sharp transient to each hit, resulting in a tighter sound, while looser snares rattle longer and therefore brighten the sound over a longer period of time.

A kick-drum tone can also be made tighter by inserting different materials inside it. While using pillows is often considered a standard, experimenting with denser materials such as sandbags can also yield excellent results. Further experimentation can be made with kick drums by using ported or un-ported resonant heads. In a rule of thumb, ported kick drums provide more attack as the sound of the beater is heard better, but sustain less than their un-ported counterparts.

6.3 The Microphones

Setting up the microphones is usually the most time-consuming yet interesting part of a recording project. A full list of the equipment available at the studio needs to be prepared (microphones, preamps, EQs, compressors), and a patching list should be created prior to the session. It is much easier to get to a studio and set up everything following your own written instructions than to work out the setup while you are addressing other issues (where are we going to source the extra cymbals? where are those shot-bags to hold the overhead boom stands? how are we going to record eighteen microphones with a sixteen-input sound card?). Coming back to the "sound map" created during the preproduction process is necessary here, as a lot of the equipment to be used should have been thought of prior to getting in the studio. The patching list should look like the following:

- Kick in > D112 > UA610 #1 > 550A #1 > 1176 #1 > Input 1
- Kick out > D6 > VT737 #1 > Input 2
- Snare top > SM57 > VHD #1 > 5013 #1 > 1176 #2 > Input 3
- etc.

When working out the different equipment combinations, try to imagine what each mic/preamp/EQ/compressor would sound like and whether or not it could give the desired outcome. While you should not commit to particular equipment before listening to it, connect everything as best as you believe it will sound, and experiment with the results. From there, experiment with the equipment chain, starting at the instrument and finishing with the ADC. This experimentation means, for example, that if the snare sound is not what you are after, the steps taken should be to do the following:

- Try different snares and tuning until the best result is obtained from the instrument
- Move the chosen microphone to different positions until the best result is obtained from the microphone
- In the event that no microphone position results in the desired sound, swap microphones and repeat the previous step
- Then move to the preamp, trying different settings there (driving the preamp, tone control, etc.)
- In the event that the preamp itself does not achieve the desired sound, swap preamps and repeat the previous step
- Tweak EQ
- Tweak compressor
- Try different ADCs

There is no point in tweaking the EQ before moving the microphone to the best possible position, just as there is no point trying a different ADC until all other options have been tried.

Experimenting with the kick-drum microphone position is essential before recording. Getting an assistant to move the microphone around while you are listening to it in the control room will work better than using the best EQs in the world to fix the kick later! This approach is especially true when you are dealing with a kick that has no sound hole (Audio 6.5 ▶). Note that if you are using two kick microphones (kick in and kick out), the best course of action is to set up the kick out microphone first using the technique previously mentioned, then flip the polarity of the kick in microphone and simply listen for the placement inside the kick drum that gives the most cancellation when added to the kick out microphone. From there, simply flipping the polarity back to normal will ensure both microphones are as much in phase as possible.

Another piece of the kit that can benefit greatly from moving the microphone is the hi-hat. Here again, moving the microphone from the bell to the edge can give very different tones. Facing the microphone down toward the bell will give a fuller and rounder tone, while placing it toward the edge of the hats will give a thinner and more hollow sound (Audio 6.6). Because of the way that odd and even harmonics radiate from the hats, a range of tones can be

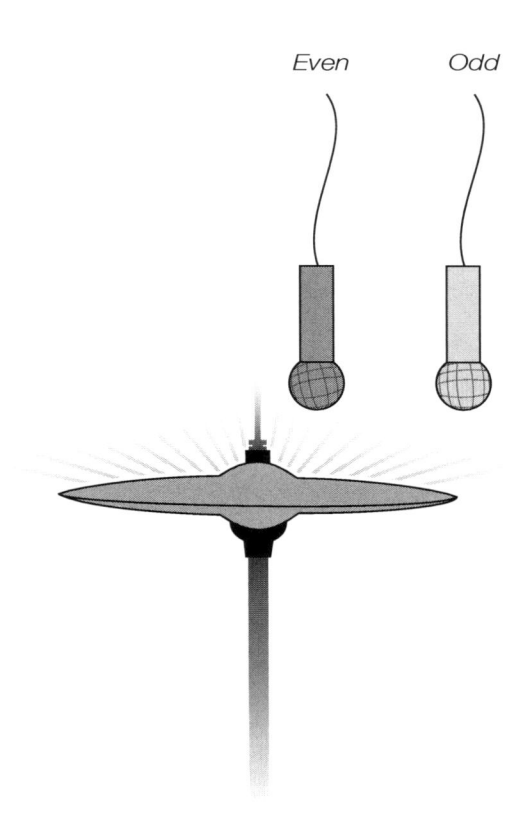

FIGURE 6.2

Hi-hat microphone placement.

achieved by moving this microphone (Figure 6.2). Be careful not to place the microphone directly to the side of the hats, as a "woof" of air will be recorded every time the drummer hits the pedal.

When you are placing a microphone on the snare drum, it is possible to control how much "crack" versus "ring" you record. A microphone placed toward the center of the drum will give more attack, while the sides will ring out more (Audio 6.7). The same concept also applies to toms. Note that when using microphones with cardioid or bidirectional polar patterns on a snare drum, it is preferable to ensure that the microphone's dead side (back of a cardioid or sides of a bidirectional) is pointed directly at the center of the hi-hats to minimize spill.

Extra ring can be achieved by ensuring that the drums are "isolated" from the ground. In much the same way that studio monitors are placed on foam or rubber mats to ensure that the low-end response is as clean as possible, placing the floor tom on a soft surface such as cymbal felts can make the tom ring longer (Audio 6.8 ▶).

When you are recording drums, it is important to remember that less is more. If a solid yet punchy kick drum can be achieved with one microphone, then this setup should be preferred over achieving the same sound with two. Try to achieve the desired sound with the least amount of equipment possible, and add microphones, EQs, compressors, and other equipment only as required. The fewer tracks in the session, the fewer phasing issues are going to creep up, and the more space the drums can occupy in the mix. Overall, always think of the simplest solution when planning the recording, and expand only if needed. One easy tip to follow when choosing the number of microphones to use on a drum kit is to listen to what the drummer will be playing on the recording, and use more microphones on the parts of the kit that are most often used. For example, if the kick-drum part rather simple, use only one kick microphone. If it is played more extensively (intricate double-kick rhythms for example), then add microphones to suit. If the snare drum is only used to keep time, playing a simple backbeat, use only one snare microphone. If there are ghost notes or the drummer hits the snare using different techniques (such as side-stick and rim-shot), then add microphones to suit. If the hats and ride cymbals are not used extensively, only use the overheads microphones on the kit. If there are more intricate rhythms played on these cymbals, then add microphones to support their sound.

A very useful microphone tip to reduce phasing issues between microphones is to have them all facing the same direction. When you are deciding on the angle of the hi-hats or toms microphones, ensuring that they are all somewhat angled toward the drummer's side of the room will help. In any case, try to ensure that no two microphones are facing each other but facing the same direction instead.

When it comes to microphones, preamps, and other equipment combinations, there is no one size fits all, and the best choice will come down to experience, availability, and musical genre. For example, using the Glyn Johns technique as opposed to close-miking every drum will give very different results, both with their own merits. There are, however, general guidelines that should be followed.

6.3.1 Room Microphones

Room microphones should be used only when the natural reverb and ambience of a good-sounding room is needed. The aim is to mic up the room, not the kit, often by using omnidirectional polar patterns. While the drummer is playing a basic rhythm, ask the assistant engineer to move a microphone around the room at different heights, in corners, and against walls, for example. You are listening for the room characteristics here, so try to note which areas give the most reverb and sustain to the drums. Pick the two that most complement each other while not sounding overly different. These areas will be panned hard left and right during mixing.

A single mono room microphone can sometimes be used as a parallel-compression track or FX channel. Placing a small-diaphragm dynamic microphone about one meter in front of the kick drum and one meter high and pointed directly at the snare should give a good "lo-fi"-sounding kit. If this channel is highly saturated and compressed (experimentation is key here), a parallel-compressed version of the overall kit can be used either to reinforce the kit's sound or as a special FX during the song. The best results here are not achieved with the best-sounding microphones, though. If you try this technique with a great condenser microphone, you may find that the sound quickly degrades to a point where it is unusable. Some good microphones to try this technique with include the Beyerdynamic M88 and Shure SM57.

6.4 The Overheads

There are a few different commonly used techniques for setting up overhead microphones. The deciding factor for using a particular technique is the purpose of the overheads. They can be used, for instance, to add width to the drums, enhance reverb, add naturalness, and add punch. It is important to note that the sound of the overheads will ultimately dictate how the kit sounds as a whole, so make sure to spend enough time getting these microphones right. While the two most common techniques are spaced pair and XY, others can give great results with a little experimentation.

6.4.1 Spaced Pair (AB)

Depending on the overall physical size of the kit, using a spaced pair of microphones can achieve a wide stereo spread (Figure 6.3; Audio 6.9 ▶). You should be careful here: small-diaphragm microphones will have the widest stereo image but yield a less natural sound when compared to the less focused large-diaphragm microphones, which tend to give a more natural feel to the kit, but with a narrower stereo spread. If the kit is small and contains only a few pieces, a spaced pair will not be able to achieve a wide image. A different technique should therefore be used.

A spaced pair of microphones used for overheads will also introduce phasing issues that need to be addressed. The two overhead microphones have to be time aligned. You can easily align them at the mixing stage by nudging one side until the snare arrives at the same time in both microphones. This adjustment will have the effect of bringing the snare to the center and making it sound fuller.

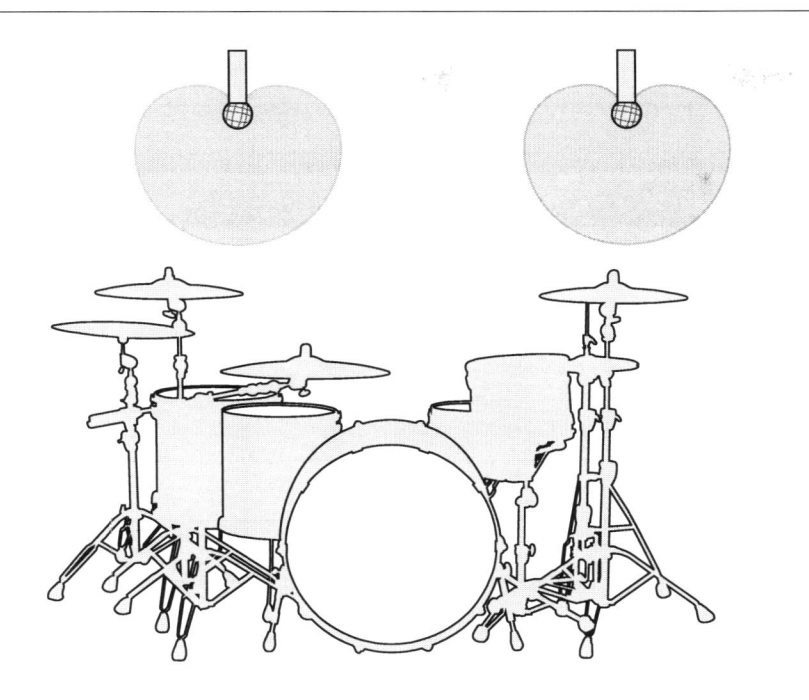

FIGURE 6.3
Overhead microphones: spaced pair.

6.4.2 Coincident Pair (XY)

An XY pair is another choice commonly used for overheads (Figure 6.4; Audio 6.10 ▶). Use this technique for the sound that it gives rather than for phase coherency. Because the drum kit is not moving while being played, the phase differences introduced by using an AB pair remain the same throughout the performance and are therefore part of the sound itself. An XY pair should be used when you are trying to achieve a more "solid" sound or when recording a kit with minimal drums and cymbals. The kit tends to be more focused, but

FIGURE 6.4
Overhead microphones: XY.

has less of a lush and wide quality when you are using an XY pair for overheads on a large kit. More often than not, small-diaphragm condensers are used for XY-type overheads, as they are more precise and directional than their large-diaphragm counterparts.

6.4.3 Coincident Pair (MS)

Using a MS pair is usually a good compromise between the lush quality of an AB pair and the focused quality of an XY pair (Figure 6.5; Audio 6.11 ▶). In this scenario, obtaining the best possible sound from the center microphone is essential, this approach could mean trialing a few different types and polar patterns. A cardioid center microphone will give the most focused sound to the kit. A bidirectional center microphone will be slightly less focused, but more open sounding. An omnidirectional center microphone will contain more low frequencies and have more of the room sound.

FIGURE 6.5
Overhead microphones: MS.

6.4.4 Coincident Pair (Blumlein)

A Blumlein pair is essentially an XY pair of bidirectional microphones angled at 90 degrees (Figure 6.6; Audio 6.12 ⏵). This technique provides some of the advantages of an XY pair such as enhanced localization of drums and strong center image while adding some of the lush qualities that a spaced pair possesses. It should be noted that this technique works best in excellent-sounding rooms with high ceilings.

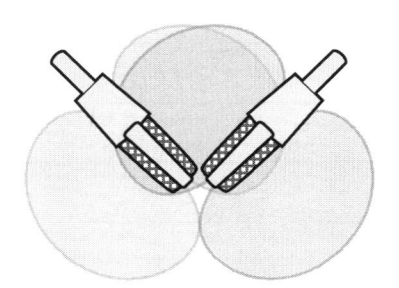

FIGURE 6.6

Overhead microphones: Blumlein.

6.4.5 Near-Coincident Pair (ORTF)

This technique, consisting of two cardioid microphones placed 17 cm apart and angled 110 degrees away from each other, works well at giving a wide and natural sound to the kit while retaining a strong center image (Figure 6.7; Audio 6.13 ⏵). It is important to note that the exact placement of the microphones can be altered to achieve a wider sound (by angling the microphones) or, on the contrary keeping a stronger center image (by bringing the microphones closer to each other).

This technique works well in rooms with low ceilings where other techniques "choke" the kit sound.

FIGURE 6.7
Overhead microphones: ORTF.

6.4.6 Glyn Johns

While this is a technique on its own (using carefully placed overheads, and spot kick/snare microphones) (Figure 6.8; Audio 6.14), it is most appropriate to mention it in this section, as only the overhead microphones differ from other more common techniques. The "overhead" microphones in this technique are placed at the same distance from the center of the snare drum. One is placed directly above the kit, slightly pushed toward the hi-hats, and angled toward the snare drum. The second microphone, still angled toward the snare drum, is placed just above the floor tom. If you use this technique, it is often possible to achieve a full kit sound with a small number of spot microphones. It is important to experiment with placement and polar patterns, as those factors can greatly affect the overall sound. Using cardioid microphones will give a stronger kit sound, while omnidirectional microphones will give a softer and more open sound. Note that you can "mix" the relative levels of the snare, hi-hats, toms, and ride cymbal using this technique by moving the microphones along an imaginary circle which has the snare drum as its center, therefore bringing the microphones closer or further away from each part of the kit.

FIGURE 6.8

Overhead microphones: Glyn Johns.

6.4.7 Microphone Height

When you are using AB, XY, MS, ORTF, and Blumlein techniques, the height at which the overheads are placed can play an important role in the final sound of the kit. Higher placement will give an "overall" sound of the kit, while lower placement will give more cymbals. Of course, if you are using separate hi-hat and ride-cymbal microphones, lower overheads may be overkill. If you place the overheads too high, on the other hand, some of the stereo spread may be lost.

The microphones should not be placed directly above the cymbals where the drummer's stick hits. The cymbal moving up and down below the microphone will introduce volume differences and extra phasing issues between the two overhead microphones. Instead, placing the microphone above the center of rotation of the cymbal will ensure consistent volume (Figure 6.9). If you are in doubt, placing the overhead microphone in the center above the cymbal will always ensure that this issue is addressed.

Overhead microphones should be recorded on a stereo track rather than two mono tracks. This method allows for the volumes of the microphones to be matched easily by looking at the meters in your DAW. Asking the drummer to hit the snare to ensure that the level in both tracks reaches the same point is the easiest way to match levels.

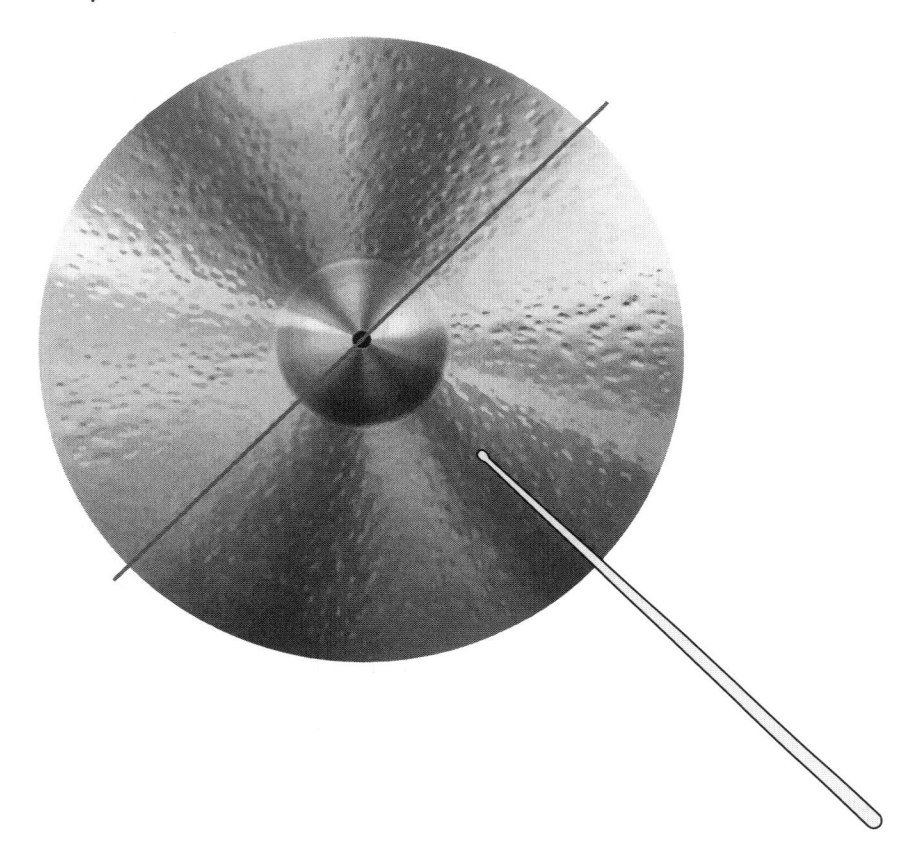

FIGURE 6.9
Center of rotation of the cymbal.

6.5 Running the Session

If at all possible, try to set the kit up and pull sounds toward the end of the day, then leave the studio and start the recording fresh the next morning. The tedious process of such chores as pulling sounds and moving microphones can bring the drummer's energy down before he or she even starts recording.

Since the question of recording to a click track or drum loop should have been answered in preproduction and, if one is to be used, the guides should have been recorded to that, whether or not to send the click to the drummer's headphones should really be a question of taste for the drummer. Good drummers often want the click, while less experienced players will ask for the guides only, as the drummers might not be used to playing with a click. This lack of experience should not be an issue if the preproduction recording was done properly. If the drummer is constantly getting out of time, it could mean that his or her headphone mix is not loud enough (drummers are notorious for needing their headphones very loud). In any case, a click is not always necessary, as modern DAWs can adapt their grid to a free-form recording instead of adapting the recording to the grid, as is the case when recording to a click.

I always recommend starting the recording session with the main single. Since the start of the session is when everyone has plenty of energy, the potential frustration of dozens of takes has not developed yet. The morale of the band is up, so getting the most important song out of the way at this point is crucial. You may want to follow it with a couple of difficult songs, since the drummer still has plenty of energy. In the event that the first recording did not go very well, follow with an easy song instead to bring the morale back up. It is the producer's job to ensure that the project runs as smoothly as possible!

Take breaks between songs to allow for the room to cool down, the drummer to rest, and the engineer to do some critical edits. Note here that if the drummer is still pumped and wants to keep going, you will have to do just that and take a break before the next song. Always remember the time left in the studio and how it can be used to maximize the outcome.

While recording, take notes of which takes were good for which parts of the song (some DAWs offer clip ratings and color coding, which work well for organizing notes). This habit will considerably speed up the editing process, as you will not need to listen to multiple takes of each part to decide which one is the best.

At the end of the recording session, sample the whole kit in case sound replacing needs to be done during mixing. A minimum of three velocity layers for each drum should be recorded, as well as recording kick and snare with different cymbals ringing out in the background. This range will give every possible combination of drums that could ever be needed should one mistake be hard to fix.

Remember to always automate the volume of the click track in the headphones to fade out toward the end of the song. There is nothing worse than hearing a click creep in as the cymbals ring out. If the click does intrude, you can always use the sampled cymbals!

6.6 Editing

Once everything has been packed up and everyone has gone home, the drums need to be edited for each song so that they are perfect for the next recording. The bass should not be recorded to a sloppy drum performance. Even if the drummer is very good and every take is perfect in terms of timing, two takes could have different kick-drum patterns, which will dictate how the bass player reacts to the performance. This point is very important: do not go back to the studio to record without having fully edited the drums, or you will only create more work and headaches later in the project.

At best, the process of drum editing should involve only grabbing all of the best takes, putting them on new tracks, and fixing minor timing issues, if necessary. If there are major timing issues or the genre calls for a near perfect performance, more drastic quantizing measures may be required. There are two different ways to quantize audio when you are recording to a click, both of which have advantages and disadvantages: cut/paste and time stretch.

6.6.1 Cut/Paste

The cut/paste method is done in three steps, and working on one section of the song at a time is recommended. The first is to separate the drums on each hit (all of the drum tracks must be grouped and separated at once to avoid phasing issues and double hits between displaced tracks). Most genres require the drums to be separated only whenever a kick or snare hit is present, as variations in timing of hi-hats can often help in retaining the life of the performance. Drum fills should not be separated on each hit but instead treated as one single event. Hard-quantized fills sound too robotic for most genres.

Next, each hit should be moved to its intended position on the grid. In order to retain some of the life of the playing, quantizing every hit 100% on the grid is not recommended. Instead, moving hits between 70% and 90% closer to the right position will keep the original feel while tightening the performance. The amount of quantization done will ultimately be dictated by the original performance and the genre of music. It is also possible to leave untouched those hits that are within a certain percentage of being right. This strategy can also be useful for keeping a more natural performance.

The final task is to fill the gaps between audio clips that have been moved away from each other and add crossfades to smooth out edits. Ensure that it is the later clip that is trimmed back in time instead of the earlier clip trimmed forward, as trimming back adds more "sustain" to the early hit rather than creating a double hit at the start of the next. Once you have done so, the section of the song that has been quantized should be auditioned for double hits, cymbal volume dips, and general feel.

In Pro Tools, a tool called Beat Detective can achieve all of these tasks rather quickly—this brand is often the DAW of choice for acoustic drum editing. The main advantage of using this method instead of time stretching is that the audio is moved only in time; therefore, the full audio quality remains. The downside of this technique is that it is more time consuming than time stretching is.

6.6.2 Time Stretching

Time stretching is available in most modern DAWs and is usually easy and quick to set up. It involves analyzing the drums as a group, creating markers for the most important transient events, and then quantizing the clips.

The quality of this method depends on the algorithm used, and the amount of stretching that needs to be done. In general, this method works well for performances that require only minor movement, as the artifacts associated are minimal. For performances that need large amounts of time stretching, this method is often not clean enough for professional productions. The main advantage of this technique is the speed at which it can be set up and the ability to quickly audition different quantizing parameters, including specific grooves.

The different algorithms available are usually made for monophonic instruments, polyphonic instruments, transient heavy instruments, complex program, and tape emulation. While they all have their own set of advantages, transient heavy instruments algorithms usually work best on drums, as the algorithms tend to stretch the end of each hit rather than the start. This stretching ultimately adds sustain or, in some cases, silence at the end of each hit.

6.6.3 Finalizing Edits

Once drum editing has been done, save a new version of the session, consolidate all the drum tracks, and move on. This method has the purpose of clearing the displayed edits in the session, as well as committing to those edits (with the ability to revert to them if necessary). Keep unused takes until the end of the project, as there may be a need to retrieve parts from other takes. Storage space is so cheap these days that it would be foolish not to. If the session needs to be sent to a mix engineer or archived after the release, delete all the extra playlists and send only the final edited version, to save on internet transfer time and ensure that the mix engineer uses the properly edited version.

6.7 Mixing Tips

6.7.1 Phase Alignment

With modern production tools, aligning the phase of the different kick microphones together, different snare microphones together, different toms microphones together, and all of these parts to the overheads is rather easy. This method will help get the best out of the recorded drums, as the comb filtering introduced by phase misalignment is reduced. When you are aligning microphones to the overheads, the snare drum is the instrument you need to deal with in priority. Since overheads can be brought down in volume during tom fills and do not contain much kick signal, time-aligning the overheads with the snare is often all that is necessary. Note that hi-hats and cymbals are not usually problematic when it comes to phasing. While most phase-alignment tools work by moving one track in time with the other (most tracks are usually aligned to the overheads), some plug-ins allow for phase alignment of some frequencies only, a technique that gives tremendous control over the sound of the kit. Do not skip this crucial first step when mixing drums, as it is more powerful than any EQ in getting the right drum sound (Audio 6.15 ▶).

6.7.2 Starting with Overheads

Since the overall tone of the kit is given by the overhead microphones, it is important to use them as the starting point when you are balancing the whole kit. Use the spot microphones only to reinforce the individual elements of the kit as required. The mistake that most beginner engineers make is to start with the kick and snare, and build around them. By starting with the overheads, you will find it easier to figure out what is missing. Does the kick lack weight? Raise the kick out mic. Does the snare need brightening? Raise the snare bottom mic. And so on.

6.7.3 Panning Cymbals and Toms

When panning hi-hat, ride, and tom tracks, listen to the overheads in solo and try to match the position of each drum. This approach will give the most natural spread, because the single close-miked instruments are placed in the same position as they are in the overhead microphones. An example of using this technique to pan a tom would be to listen to the overheads in solo, listen to where the tom should be, pan the tom while you are still listening only to the overheads, un-mute the tom, and listen for a shift in position. If the tom moves in the stereo field when it is added, mute it, adjust its position, and repeat the process until the tom does not move whether it is muted or un-muted (Audio 6.16 ▶). Note that while the decision to pan "drummer's perspective" or "audience's perspective" is largely a creative one, it is preferable to pan by using drummer's perspective for genres in which listeners are likely to play air drums.

6.7.4 Kick Drum Weight

The "push-pull trick" of Pultec equalizers can be reproduced with a digital EQ by using resonant shelving filters (Figure 6.10; Audio 6.17 ▶). The stock Pro Tools EQ can achieve this shape by using its shelving filters. The trick is to find the resonant frequency of the kick drum (hopefully, the root note of the song or the perfect fifth) and push it up. This technique works better in parallel than as an insert on the kick track.

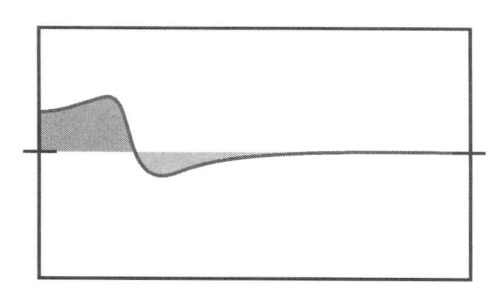

FIGURE 6.10
"Push-pull" equalizer curve.

6.7.5 De-esser on Kick

De-essers can decrease the attack or the sustain of a kick drum (Figure 6.11; Audio 6.18 ▶). For example, if the sustain needs to be reduced, set up the de-esser to be triggered by low frequencies. This step will effectively let the transient through and compress the low-end sustain. On the other hand, if you want to reduce the attack but let the sustain part ring out, set up the compressor to react to high frequencies instead. This technique is very useful when you are layering kicks and often gives better results than equalizing.

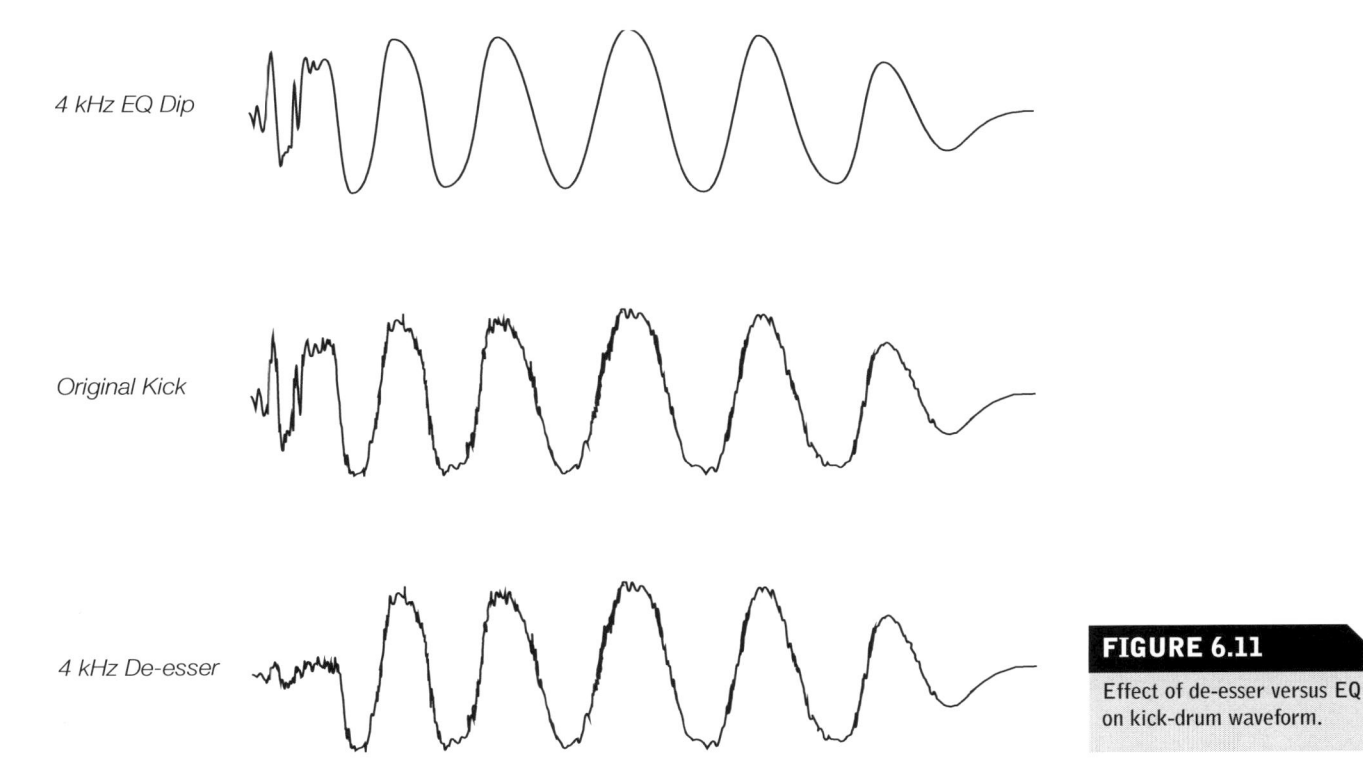

4 kHz EQ Dip

Original Kick

4 kHz De-esser

FIGURE 6.11
Effect of de-esser versus EQ on kick-drum waveform.

6.7.6 Kick Drum Attack

It is possible to increase the attack of a kick drum by using upward expansion. Using a multiband compressor acting only on the mids (around 4 kHz) with a ratio lower than 1:1 will effectively increase the volume of the transient without affecting the overall tone or sustain (Audio 6.19 ▶).

6.7.7 Normalizing Kick Drums

Using functions such as Pro Tools' Beat Detective, separate every kick hit, then normalize each clip individually. This step will help achieve consistency of volume, which is very important for this instrument. Since drums give different tones when hit differently, normalizing volumes will not make them stale, as the differences in tone will remain.

6.7.8 Sound Replacing Snare Drums

Depending on the genre of music, consistency of volume may also be required on the snare track. Sound replacing the snare usually works better than the normalizing trick. Because of the amount of hi-hat and cymbal bleed on the snare track, using the normalize function will create distracting volume changes in those instruments.

6.7.9 Bottom Snare Microphone

It is quite common to use a microphone below the snare drum to capture the rattle of the snares. While audio engineering students usually know about this technique, they often misunderstand its use at the mixing stage. This microphone has two uses: giving definition to the snare where intricate rhythms and ghost notes are played, and to brighten up an otherwise dull snare drum. Just as some engineers sometimes side-chain-trigger white noise from the snare drum to brighten up the snare, this bottom microphone should be pushed up until the right tone has been achieved. Its aim is not to have a sound of its own, but rather "support" the snare drum (Audio 6.20 ▶). It is preferable to think of this track in its supporting role as a replacement for a snare EQ (top-end boost only) rather than being its own track.

6.7.10 Recording the Snare Air Hole

You can achieve a different snare sound by placing a microphone directly in front of the air hole (the hole on the side of the shell, which is used to equalize the pressure in the drum when it is being hit) (Audio 6.21 ▶). Since there is a large amount of air being pushed from there, the microphone will record more low-end sustain with every hit. This microphone will also sound like a good blend of top and bottom microphones. Note that it is sometimes necessary to create an "air tunnel" between the hole and the microphone by using tape. More often than not, dynamic microphones will be more appropriate for this technique because of their higher SPL handling quality.

6.7.11 Gating Toms

For songs in which the toms are used only for fills, manually deleting the audio on the tom tracks when they are not being hit is the best way of ensuring that the cleanest tom sounds remain. In such songs, gates are not necessary on tom tracks. When you are dealing with songs in which the toms are an integral part of the performance, gates and expanders can be useful in cleaning up the rumble and adding punch (Figure 6.12; Audio 6.22 ▶). In this case, the attack setting needs to be fast enough for the gate to be fully opened at the start of the hit (look-ahead is useful in this case). The release needs to work "with" the rhythm and slightly shorten the length of each hit, making the toms sound a little thinner. From there, using the EQ technique described in section 6.7.12 will re-inject thickness in the tom sound.

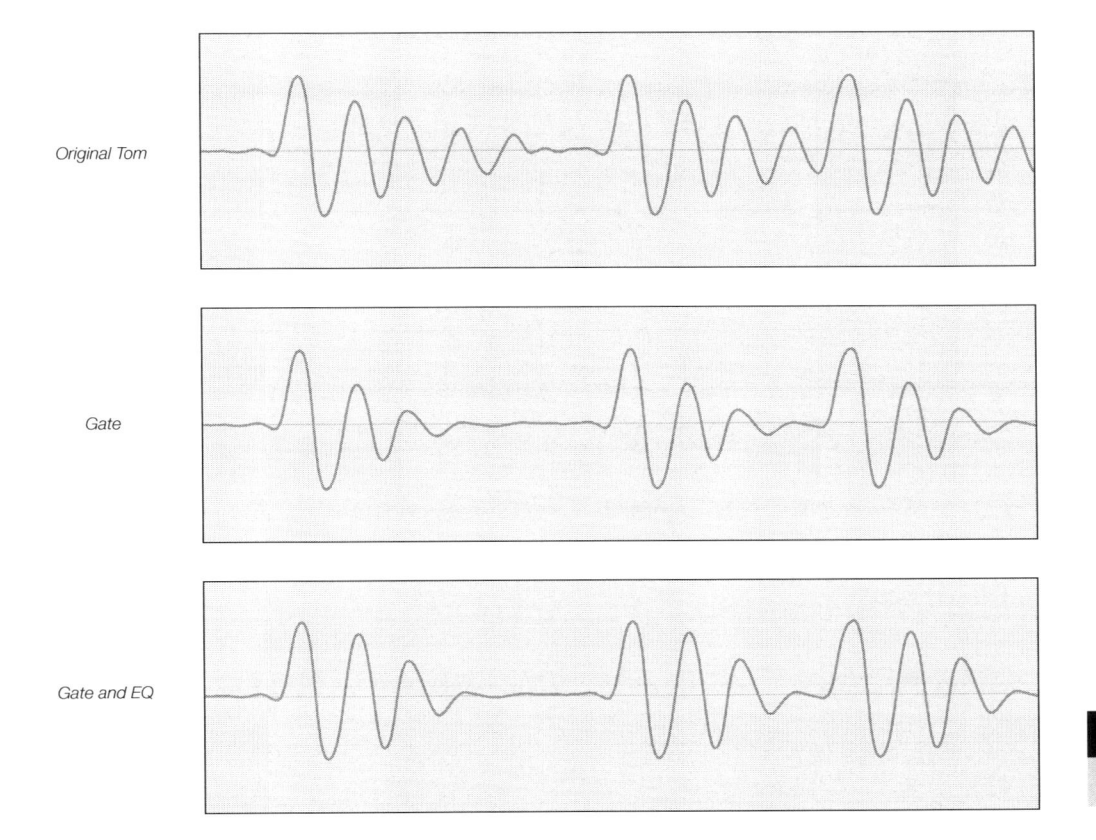

FIGURE 6.12
Gate and EQ on toms.

6.7.12 Tom Equalizing

An easy way to equalize toms is to sweep with a fairly sharp bell boost in the low end until the resonant frequency of the drum is found (the EQ will overload and distort when you have found it). Boosting a couple of dBs at that frequency, then using a steep high-pass filter just below it and a fairly wide bell cut just above will give a large and warm sounding tom (Figure 6.13; Audio 6.23 ▶).

It is often also useful to slightly dip the low mid-frequencies just above the resonant frequency of the tom to ensure that it does not sound boxy or muddy. Of course, this technique may not suit busy tom rhythms, but works well for tom fills where size and sustain is required.

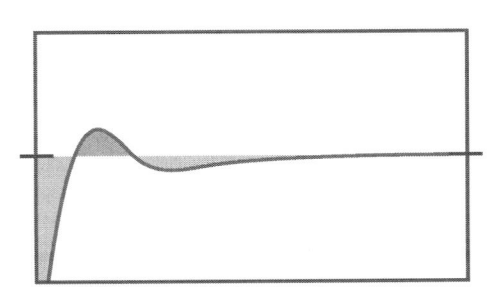

FIGURE 6.13
Tom EQ curve.

6.7.13 Hi-hat Compression

Hi-hats often allow for heavy compression. Achieving a healthy amount of gain reduction tends to increase the overall size of the kit, as the hats sustain better and therefore fill more frequencies over time (Audio 6.24 ▶).

6.7.14 Out-of-Tune Kick and Hi-hats

Since the human hearing requires more time to establish pitch at extreme low and high frequencies, it is sometimes possible to "hide" an out-of-tune kick drum or closed hi-hat by shortening their length. Using transient designers to reduce the sustain of these instruments can work well in this case (Audio 6.25 ▶).

6.7.15 Loud Drums in Low Dynamic Range Genres

Some genres of music require instruments to be loud all the time. If the drums need to remain loud, use hard clippers or brick-wall limiters to increase the perceived size of the drums (Figure 6.14; Audio 6.26 ▶). Since hard clippers effectively square off transients, and square waves contain extra odd harmonics, the drums will sound fatter and louder at the same peak level.

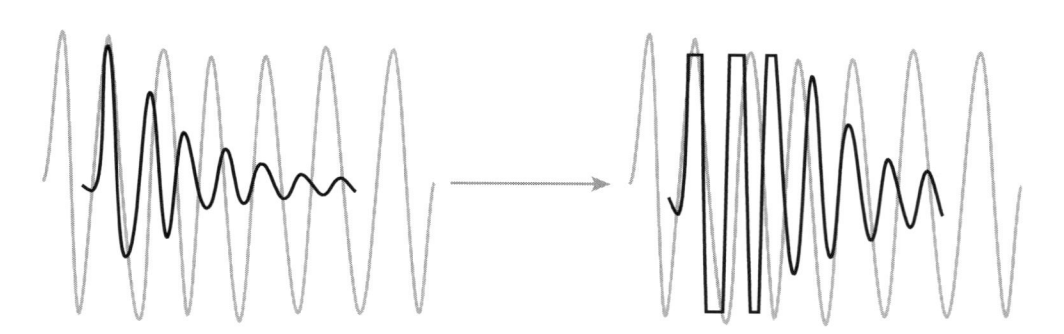

FIGURE 6.14
Effect of hard clipper on drums.

6.7.16 Transient Designers

Rather than reaching for compressors to make drums punchier, try using transient designers (Figure 6.15; Audio 6.27 ▶). They have a way of giving very sharp attacks to drums in a sound a little different from that of compressors. Excellent results can also be achieved when you follow this process with tape saturation.

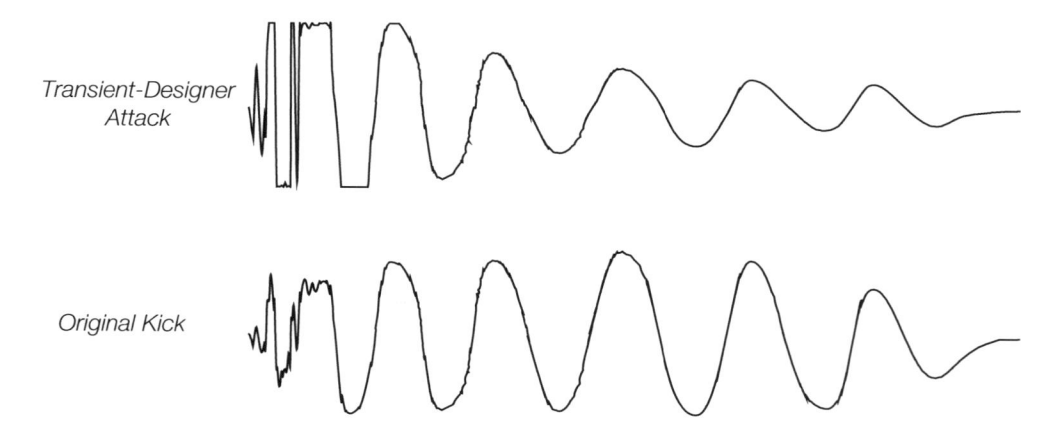

Transient-Designer Attack

Original Kick

FIGURE 6.15
Effect of transient designer on kick-drum waveform.

6.7.17 Double-Tracked Drums

One way to make the drums sound "older" or more lo-fi is to double-track the performance (Audio 6.28 ▶). Just as double-tracking guitars or voice blurs the image of these instruments, the drums will have less punch and may fade in the background a little better. When you are using this technique, it is sometimes important to quantize the performances so that the hits are not too far off each other, so as to not draw attention to the technique.

6.8 Creating Interest with Drums

6.8.1 Swapping Parts of the Kit

The parts of the drum kit can be changed for different parts of the song. For example, the first half of a verse could use one snare, while the second half could have a different one. The cymbal position can also be changed during the bridge to give the drummer a different feel for his or her own kit, and therefore play differently. Another obvious choice would be to use different stick types during parts of the song (mallets for intro, brushes for the first half of the verse, hot rods for the second half, sticks for the chorus, etc.). Some engineers even go to the extent of using different kick drums in different sections of songs!

6.8.2 Extra Microphones

Set up any microphone in the corridor/cupboard/in a box in the drum room, and record it onto a spare channel when you are recording the drums, to give another choice when you are faced with an uninteresting part of a mix. This technique is sometimes used, for example, in pre-choruses, the last bar of a chorus, and the first four bars of a verse. It does not cost anything and can help create interest at the mixing stage. A deviated technique here would be to set up two different sets of overhead techniques (XY and spaced pair, for example), and using only one at any given time, for a good way of having the lush characteristics of a spaced pair during the verses, then swapping to the strong and focused sound of an XY pair during the choruses. This mixing of microphones can be pushed even further by muting all microphones but the room mics during a pre-chorus or a bridge, giving a live feel to the part, before returning to a tighter sound. The overall aim with microphones on the kit is to keep the interest of the listeners going by fooling them into thinking that different kits were used during the performance.

6.8.3 Micro-modulations

Using micro-modulations on parts of the kit can work wonders in genres of music in which drums sound the same from start to end of the song. Some examples of micro-modulations that work well for drum kits are the following:

- Slow flanger on hi-hats or overheads, causing the high-frequency content to have movement throughout

- Low-frequency-oscillator-modulated filters on snare, resulting in different brightness and power for different hits
- Slow automation of reverb on the snare to give slightly different tones to each hit

6.8.4 Letting the Drummer Loose

While more of a production decision rather than an engineering one, the choice to have the drummer play simple grooves or really let loose can be used to create interest in the song. For example, once most of the takes have been done, it could be worth asking the drummer to record one more take while completely "over-playing" his part. The results could give a few good options for extra fills if needed, or random ghost hits, for example. Pushing this concept further, you could ask the drummer to go for a run around the building and go straight into the recording room to record a chorus. This diversion may influence his or her performance by building adrenaline and could ultimately create a sound different from that of the cleaner takes. Overall, there are no rules here; the aim is to create randomness in the performance to allow for a more interesting mix down the line.

Chapter 7

Bass Guitar

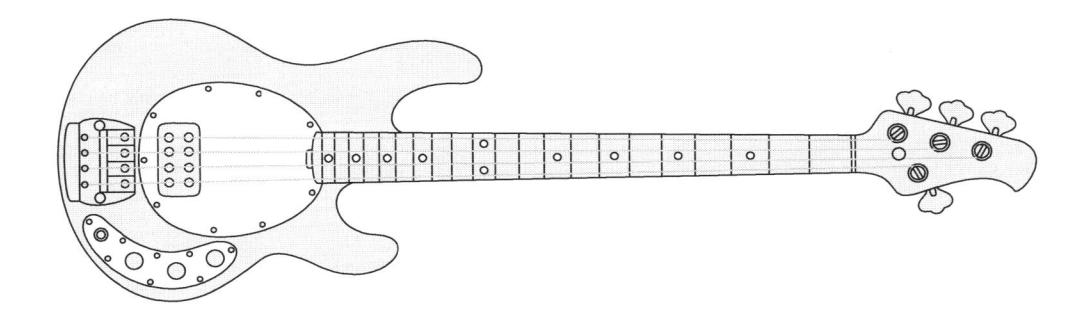

Bass guitar is usually the easiest instrument to record: the range of tones available is limited. Nonetheless, the same methods of pulling sounds apply, as there is a tendency for inexperienced engineers to brush over this crucial part of the recording. It is advisable to remember that a strong song starts with a strong rhythmic foundation!

7.1 The Room

Unless a special effect is required, recording a dry tone for the bass is preferable. The low end can get very muddy with only a small number of effects. When you are recording in an over-reverberant room, the recording may also contain unwanted inconsistencies and phase cancellations. There are, of course, exceptions to this rule, and the vision should dictate how to record. Try to use the room to your advantage. For example, when you are working with a small amplifier, placing it in a corner will make use of the bass build up. On the other hand, if the amp is rather large, moving it away from the walls and facing an open area of the room could help clean up the recording.

7.2 The Instrument

As always, the choice of instrument is key in obtaining the right tone, and once again, so is preproduction. In this case, both the bass and amp are "the instrument," and the right combination must be chosen. A Fender Jazz Bass running through an Ampeg SVT amp will sound completely different from a Music Man Stingray going through a Gallien-Krueger RB amp. The bass player will usually have his or her preferred bass and amplifier combo, but the engineer has to decide whether it is the best combination for the required outcome. As mentioned earlier, get the bass sounding right in the room by first tweaking the settings on the instrument, then the settings on the amp. The process in this case is to "reset" all the settings on both the bass and amp, then while playing at a healthy volume level (without driving the amp), start experimenting with the different pickups and tone controls on the bass itself until the best sound is achieved. Only then should you experiment with amp settings such as EQ and overdrive. After the best possible tone is achieved, it is time to select the best microphones to match it.

7.3 The Microphones

Using a minimalist approach when it comes to bass cabinets is often preferable. A single microphone well placed will always give better results than a number of microphones put up anywhere in front of the cab. Where this single microphone should be placed will be dictated by the area where the bass sounds the fullest and punchiest in the low mids. Walk around the room as the bass player is performing his or her part and note where the bass sounds best. After that, bring up the two microphones that could best complement the sound, and place them both there. A quick listen through the control-room speakers should tell you straight away which microphone should be kept. At this point, try placing the remaining microphone in different positions before completely discarding it. The second position is often a couple of centimeters in front of the cab. Only if both microphones sound great should they be kept. Otherwise, keep only the best-sounding mic for the job.

Try to avoid making judgments on the quality of the sound on the basis of sub-frequencies. Focusing your listening to the low mids and effective bass usually yields better results, since a lot of sub-frequencies will be cut at the mixing stage. The angle of the microphone is also of little importance for this instrument, as bass frequencies are omnidirectional.

DI boxes—use them! It does not cost anything to use a DI box, which can give a lot of choices at the mixing stage when the box is used with bass-amp simulators (Figure 7.1). While guitar-amp simulators tend to sound somewhat "digital," bass-amp simulators can work wonders in fooling the listener. It is obvious that a high-quality DI box is best, but any DI is better than no DI. These boxes are also useful when you are dealing with bass players who insist on using FX pedals because it is "part of their sound." No matter the genre, unless the bass is playing a solo, a decent amount of clean low end will be needed to carry the mix. A DI can save you from having to mix the song with a weak low end because the bass has a chorus FX the whole way through.

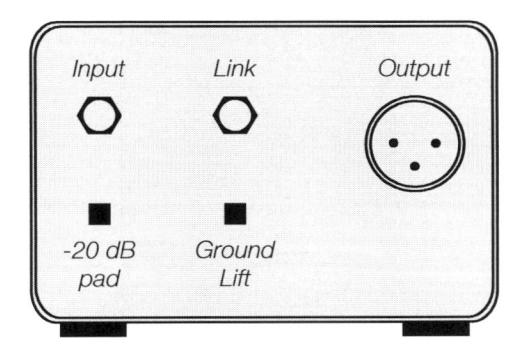

FIGURE 7.1

Direct-input box.

7.4 The Processors

Preamps can greatly affect the tone of a bass guitar, so experiment before deciding which one to use. While each preamp is being tried, make sure to drive the preamp, since the various models react differently when overdriven. You might find that your best preamp sounds good when it is clean, but that a less expensive one sounds even better when driven a little. Experimentation is vital! The bass often gets less attention than other instruments, but make sure to spend the little time necessary to get the required tone. A great mix starts with tight low end.

7.5 Running the Session

Since the drums have already been edited, the bass can be recorded to a solid rhythmic foundation. Songs rarely have two bass lines playing at the same time, and this instrument is simple enough that buzzes and other mistakes can be heard during recording. Thus all "tone" edits can be done on the spot. Do not record this instruments on different playlists; overwrite any part that is not good enough. Working this way will not only save time after the session is finished, but it will also ensure that the right performance is recorded.

It is often preferable to have the bass player perform in the control room. It should only involve running the bass through the studio tie lines or, if tie lines are not available, using a slightly longer tip-sleeve (TS) cable or using a DI box as an extension. Having the musician in the control room has quite a few advantages, and since bass players never interact with their amp for feedback or tone, there are no real benefits to having them play in the live room. The first and most obvious advantage is a clear communication path between engineer and performer. It is easier for the bass player to try an idea or the engineer to show how a particular edit would sound when they are both in the control room. Once the settings on the amplifier have been agreed on, having the musician away from the live room ensures that he or she will not modify the setup. Since a lot of energy can be obtained by running a loud mix through the studio monitors, the performance may improve as well. This result is, of course, dependent of the monitors available in the control room, as far fields are great for this purpose but small near fields may not work as well. Also, performers who record in the control room ask to listen to their performance more often. Hearing their music as it is recorded and edited can be a great way for bass players to hear their performance and therefore work out what could be improved on the spot.

Always ask that bass players tune perfectly at the start of each song. Different players tune their instruments differently. Some go a little sharp to allow for long sustained notes to sound right, while others tune perfectly for small notes, making longer ones go flat.

7.6 Mixing Tips

7.6.1 Quantizing Bass

It is very important to quantize bass to the groove given by the kick drum (Figure 7.2; Audio 7.1 ▶). If the drums have been recorded and edited first, the process should be very quick. Groove extraction is available in most DAWs and should be used on every song. A tight low end given by the kick and bass is the foundation of a great mix. Note that the cut/paste technique is preferable over time stretching in this case to ensure that the bass sound is kept intact.

Unquantized Bass

Kick

Quantized Bass

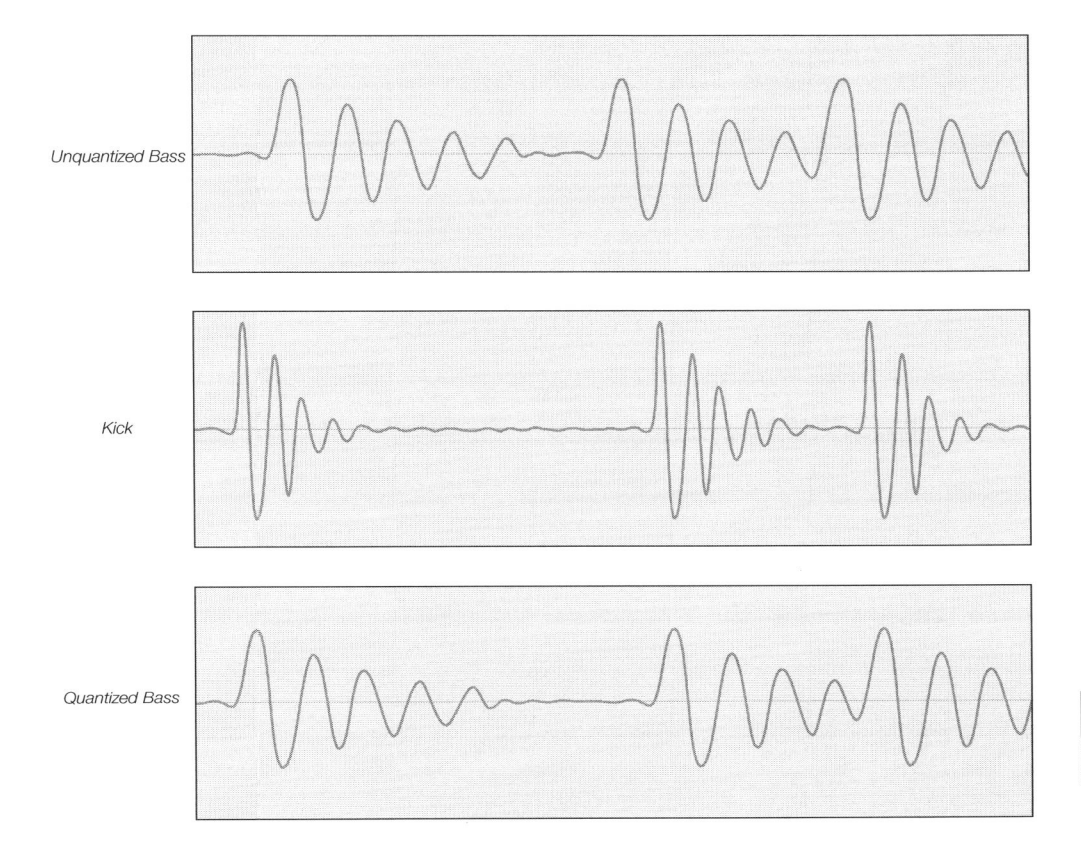

FIGURE 7.2

Quantized bass guitar.

7.6.2 Kick and Bass Relationship

It is important to take into account the duration and frequency content of both kick drum and bass to ensure that the best separation is achieved between these two instruments. A long kick drum that contains a lot of sub-frequencies should be matched with a bass that has been high-pass filtered or plays rather short notes. A short kick drum can be matched with a bass that extends lower and plays longer notes (Audio 7.2 ▶).

7.6.3 Heavy Compression

Compressors can be pushed harder on bass instruments: they can turn weak bass tones into heavier and fatter ones. Be careful not to set the release of the compressor too short, as it can introduce unwanted distortion. The volume of the bass needs to be consistent (using automation works best) before compressors and limiters are used. Bass tracks often have resonant frequencies caused either by the instrument itself, the room it is recorded in, or an untreated mixing room. To know which notes should be turned up or down, put a root-mean-squared (RMS) meter on the bass and ride the levels so that the meter is always hovering around the same value. If you add a compressor and a limiter in series after the volume automation, the bass will not jump out on any particular notes. Depending on the type of compressor being used and its detection circuit, some notes may be heavily compressed, while others are left intact, resulting in different tones for different notes. These varying results are why riding the volume before using the compressor is so important.

7.6.4 Thickening Bass

A good way to thicken a bass guitar is to saturate it (Audio 7.3 ▶). Depending on the type of saturation or distortion being used, some of the low end may disappear, and the top end may become harsh. If you get these unwanted results, it is often advisable to use the saturation in parallel in order to keep the low end of the original bass, and apply a gentle high- and low-pass filter on the saturated signal to smooth out the distortion.

Another way to thicken the bass is to add vibrato to it (Figure 7.3; Audio 7.4 ▶). A slight vibrato effect applied to the bass track can help fatten up the low end as the bass will occupy a larger frequency range for each note played.

FIGURE 7.3
Bass vibrato.

7.6.5 Sub-bass Enhancement

Some genres of music may require the bass to extend lower in the frequency spectrum than the original recording. There are two main techniques used to achieve this requirement.

This first and simplest way to extend the range of a bass instrument is to use sub-bass enhancement processors (Audio 7.5 ▶). Over the years, many hardware processors and plug-ins such as the Aphex Big Bottom or Waves LoAir have managed to successfully achieve this enhancement. Their inner working is rather simple: they track the pitch of the bass and generate a synthesized sine wave an octave lower than the original. The result is a sound that retains all the tonal qualities of the original bass while lowering its fundamental frequency. When you are using these processors, it is often necessary to check that no phase cancellation is introduced between the original and synthesized bass on a monitoring system that can reproduce sub-frequencies.

The second technique, while more time consuming to set up, can yield better results, as it allows finer control over the final bass sound. It involves converting the bass line from audio to MIDI, and using a synthesizer to play back the subharmonics. While this process was hard to achieve in the past, modern DAWs and plug-ins can easily and quickly give usable results. Once the audio has been converted to MIDI, experimentation is critical in order to find the best synthesized bass for the song.

7.7 Creating Interest with Bass

Bass is one of the only instruments that need to keep steady and unchanged throughout a song. Since it carries most of the "weight" of the song, having dramatic changes can result in a song that feels weak in certain parts, and too strong in others. While it should be done carefully, using EQ, distortion, and micro-modulations can help create interest. The rule is to split the bass track into lows and highs by using high- and low-pass filters of linear-phase equalizers. Two hundred hertz is often a good area to split the bass, as everything below is mostly sub-frequencies and effective bass. Once you have split the track, modulating the high-frequency content can give movement to the bass without affecting its strength. Effects such as evolving chorus or distortion can work well in the low mids.

Chapter **8**

Electric Guitar

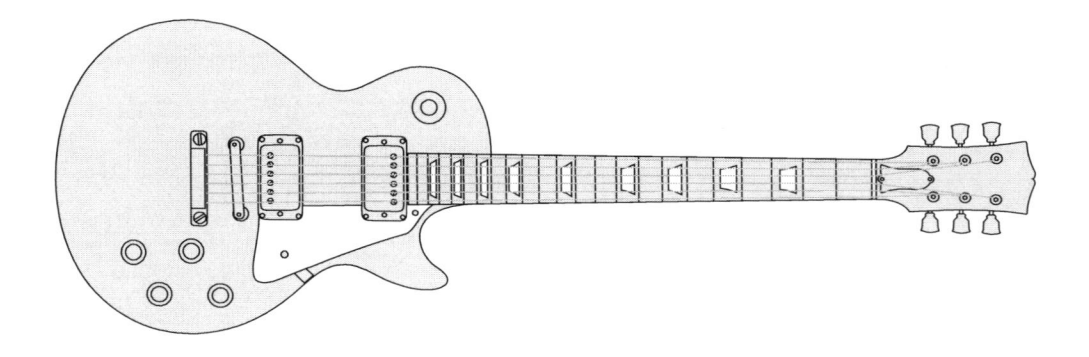

Electric guitars can offer a large range of tones and steer the recording toward a particular genre. While the recording process is not overly complex, experimentation is necessary in order to achieve the best possible sound. This chapter offers insights into such topics as the different guitar models available, strings, and pickup types. While these specifics may not seem relevant to engineers at first, it is important to remember that large recording studios often have a range of guitars available to use, and just as an engineer needs to be able to pick the right microphone for an instrument, modifying the sound source itself (the guitar and amp, in this case) is just as important in order to achieve a great tone. Even if a guitarist is set in using his or her own equipment, being able to "talk the lingo" with the musician can help build a rapport in the studio and ultimately capture better performances.

8.1 The Room

While not nearly as important as the room in which drums are recorded, the guitar cab calls for some guidelines to follow during recording. For this instrument, the room itself is often not as important as the position of the cab within the room. Placing the cab on the floor creates a bass buildup that microphones will pick up. This buildup could be useful when there is no low-end instrument in the song, in which case the guitar will have to carry the low-frequency information. If the song already contains a bass guitar, lifting the cab off the floor will ensure that the recorded sound does not contain any unwanted low-frequency content that could potentially add mud and clutter the mix. This technique is used by a lot of engineers who place guitar cabs on chairs.

The farther back the microphones are placed from the cab, the more room ambience will be picked up. This combination has the effect of pushing the recorded sound further back in the mix, a result that could be particularly useful when you are working with a cluttered arrangement.

8.2 The Instrument

Knowledge of the tones that can be achieved with the main guitar designs such as the Fender Stratocaster, Fender Telecaster, and Gibson Les Paul and amplifier brands such as Vox, Orange, and Marshall will greatly speed up the process of achieving the right tone for each part of the song. A Telecaster running through an Orange amplifier will allow for completely different tones from those of a Les Paul running through a Mesa amp. Engineers also deliberately mismatch amp heads and cabs or connect two cabs from the same head, thus greatly expanding the range of tonal possibilities. Experience is key here, so the choice ultimately relies on having listened to a range of models, and making an informed decision on the assumed best combination. There are, however, some rules that apply to all brands and models when it comes to strings, pickups, and general settings.

The choice of TS cable is also quite important. Investing in a good-quality guitar cable that is used only for recordings is essential for audio engineers. Never let the guitarist record with the same lead he or she used at a gig the night before!

8.2.1 Guitar

8.2.1.1 Strings

Light-gauge strings on guitar usually equate to a thinner sound, while heavier gauge will yield fuller tones (Figure 8.1). Ultimately, the guitar player will choose strings that he or she feels comfortable playing, but the engineer needs to be aware of what is being used to accurately predict whether or not the final sound will work with the vision of the song.

FIGURE 8.1 Guitar string.

8.2.1.2 Pickups

There are two main pickup designs used in electric guitars, both of which have their advantages:

- Single-coil pickups can introduce noise, but have a more "precise" sound (Figure 8.2; Audio 8.1 ▶). They are often used for lead lines, as they tend to cut through mixes better than the second kind (humbuckers).

FIGURE 8.2 Single-coil pickup.

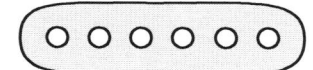

- Humbucker pickups are noiseless, and are generally considered to have a warmer or richer tonal quality (Figure 8.3; Audio 8.2 ▶).

FIGURE 8.3 Humbucker pickup.

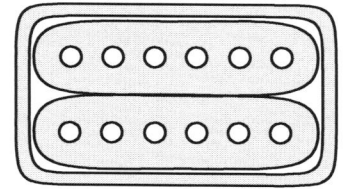

Pickups can also be passive or active, both of which have advantages and disadvantages:

- Passive pickups have a wider dynamic range, which allows for more dynamic playing, but the tone control is limited to a simple

low-pass filter, the pickups have a lower output, and, in the case of single coil pickups, they are subject to noise and hum.
- Active pickups are more consistent in volume, have a higher output, are not subject to noise and hum, and can have a wider range of tone options available than passive pickups. The downside is that dynamic playing is harder due to the consistent volume output, and the pickups require a battery to work.

8.2.1.3 Pickup Position

These pickups can be placed toward the following:

- The bridge for a more "plucky" or "defined" sound (Figure 8.4; Audio 8.3 ▶). Usually the first choice for leads.

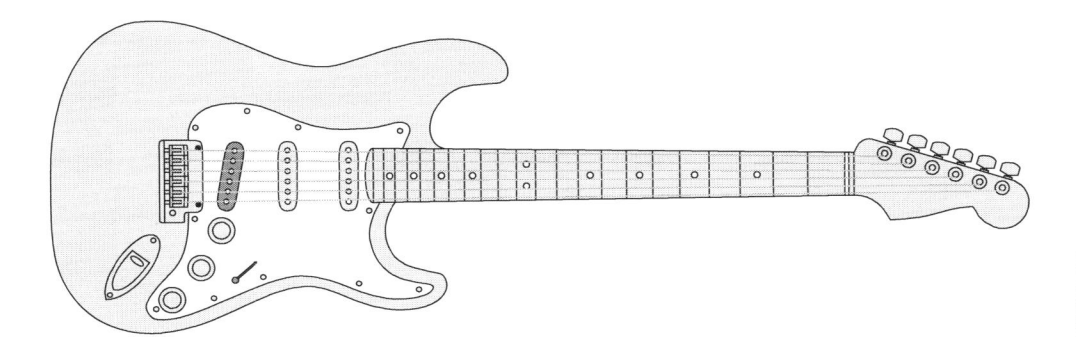

FIGURE 8.4
Bridge pickup.

- The neck for a sound with less attack, or a smoother sound (Figure 8.5; Audio 8.4 ▶). Usually the first choice for rhythm guitar.

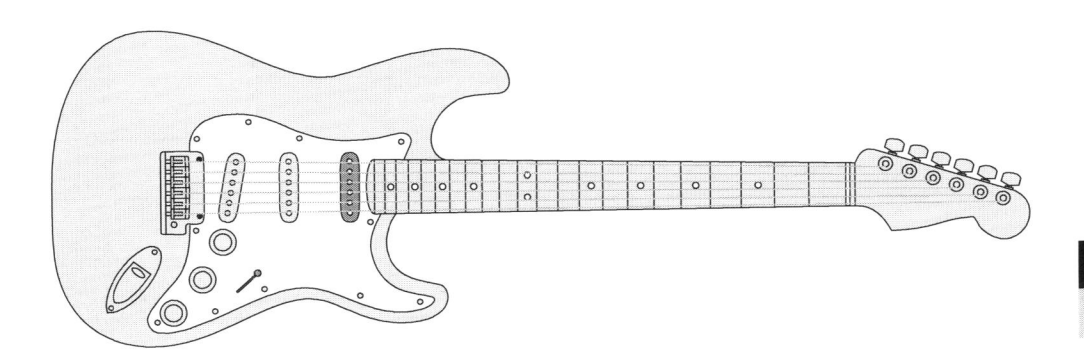

FIGURE 8.5
Neck pickup.

Some guitars offer the choice of putting the pickups out of phase with one another (Figure 8.6; Audio 8.5 ▶). Doing so results in a "hollow" or less "present" sound that blends well with other midrange-heavy instruments. While this kind of pickup is not often the first choice for lead guitar parts, it can be useful to make a rhythm part fit within a busy mix.

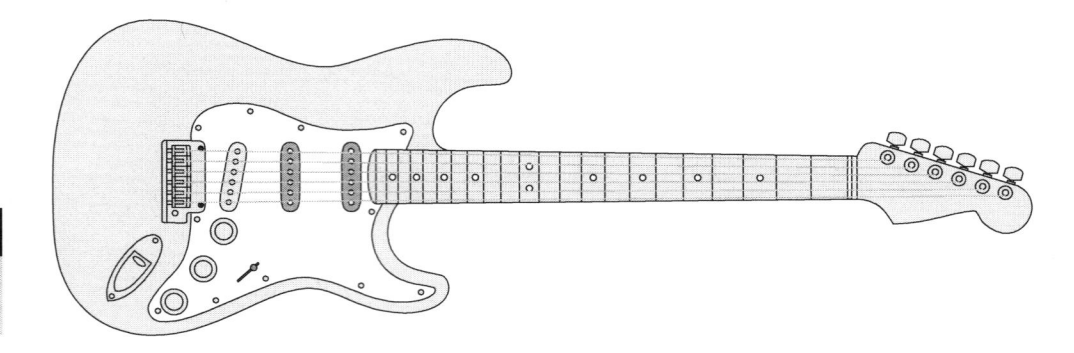

FIGURE 8.6
Out-of-phase neck and middle pickups.

8.2.1.4 *Volume Knob*

The volume knob on electric guitars serves two purposes: to raise the overall volume of the instrument going into the amp, and to saturate the sound (Figure 8.7; Audio 8.6). Depending on the settings, this knob will achieve either of those effects. The first half of the knob is purely a volume control. Once the knob has reached the halfway point, it becomes a saturation control, while keeping the overall volume the same. This means that at 12 o'clock, the volume will be clean and loud, compared to fully turned clockwise, which will also be loud, but with a more saturated tone.

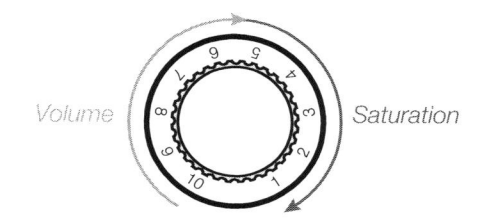

FIGURE 8.7
Guitar volume knob.

8.2.2 Amplifier

8.2.2.1 *Tube versus Solid-State Amps*

Guitar amps come in two basic flavors: tube or solid state. There are pros and cons for each design, and they should be taken into consideration when you are picking which amp the guitar will be plugged into for the recording. Tube amps tend to sound warmer and distort quite nicely when overdriven. The distortion added is similar to soft clipping, as it adds odd harmonics that drop in volume quite rapidly in the higher end of the frequency spectrum. This result is similar to the frequency content of triangle waveforms (see Chapter 11, "Synthesis Basics," for more information). The downside of using tube amps is that they are more prone to failing, are quite fragile, and are generally more expensive. Solid-state amps have a more neutral and bland sound, which can sometimes be described as cold. When overdriven, solid-state amps give a tone similar to hard clipping, which also adds odd harmonics, but is a lot harsher, as the harmonics remain high in volume even in the higher end of the frequency spectrum. This result is similar to the frequency content of square waveforms (see Chapter 11 for more information). Solid-state amps are generally more reliable and less expensive than their tube counterparts.

8.2.2.2 *Speaker Cone*

Placing a microphone in the center of the cone or to the side will affect the recording (Figure 8.8; Audio 8.7 ▶). The center position will give a brighter tone, while placing the microphone to the side will give a more mellow tone. Using either of these positions in conjunction with off-axis positions allows for a range of different tones to be achieved.

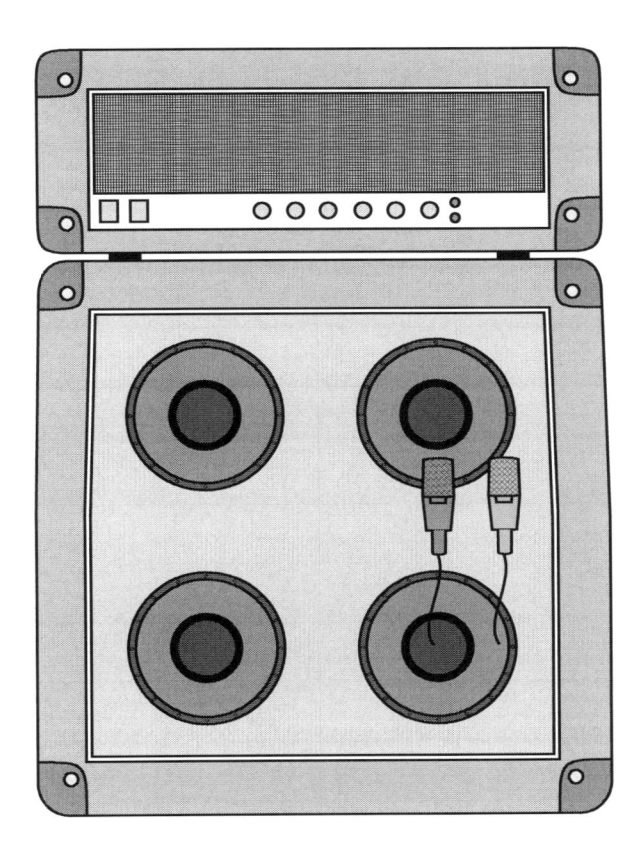

FIGURE 8.8
Guitar-cab speaker-cone recording position.

Always listen closely to each cone on the amplifier before setting up microphones. Different cones of the same cab can sound vastly different. Listening to each cone (at low volume) may reveal that, for example, one sounds piercing, another is warmer, and a third is slightly distorted. The microphone position can then be matched to achieve the desired tone (Audio 8.8 ▶).

8.3 The Microphones

There are a lot of different possible combinations when it comes to miking up a guitar cab. Some engineers use one, two, or more microphones. The microphone can be put close to or far from the cab, or behind it, for example. As always, there is no wrong way of recording as long as the end product is what the song needs. The following two techniques are the most commonly used among engineers. The third technique is one that I developed over the years; it allows for a lot of flexibility without being overly time consuming. Note that using a DI on the guitar, while often discarded at the mixing stage, can be an option if more experimentation is needed in terms of tones/amp simulation/special FX/re-amping. In this case, try to have the DI signal running through the cleanest and least colored preamp.

8.3.1 Pink-Noise Technique

This technique is the most famous one for guitar-amp recording and the one that works best in single-microphone setups. The first step is to plug a pink-noise generator in the amp and play it back at low volume. Next, having chosen the most appropriate microphone to complement the original guitar sound, listen to the microphone through headphones in the recording room. Move the microphone around and note the positions where the sound in the headphones is brightest, dullest, and fullest, and as close to the actual noise as possible. These positions will have the same effect on the guitar sound, so it will only be a matter of choosing the position that is required for the intended sound. More often than not, the position that sounds closest to the real noise will be combined with one of the others to achieve a wide range of tones.

8.3.2 Out-of-Phase Technique

This technique works well when time in the studio is limited. There are different variants of this technique that use two or three microphones. Using two microphones only, set one up anywhere in front of the cabinet and move the second while listening to both until the phasing between the two achieves the desired sound. Because of the interactions between the two microphones, a variety of tones can be achieved. Note that the positions where the microphones sound thinnest (or fullest) can achieve the opposite tone if you flip the polarity of one microphone.

Using three microphones gives more flexibility at the mixing stage. The first two microphones should be placed close to the cabinet, while the third can be placed farther back (a fourth and fifth microphones can also be used in different positions if more flexibility is needed at the mixing stage but time is very limited at the recording stage). The aim there is to record all microphones and modify the phase of the different microphones during mixing to achieve the different tones needed. While this technique can give excellent results, it is not recommended if you have enough recording time to use the other techniques, as too much of the final sound achieved is left to chance.

8.3.3 Best-Two-Microphones Technique

The final technique, and my personal favorite, requires more microphones. The first step is to put up all of the microphones that could work in front of the cab. This arrangement often means having between four and six microphones all running through the same preamp type as a starting point. While there are some commonly accepted industry-standard microphones such as the SM57, MD421, C414, and U67, any microphone could potentially work when you use this technique. Once all of the microphones are ready, record a four-bar loop of the song, time-align all the tracks in the DAW, and then carefully audition all two-microphone combinations possible. Listen for the two microphones that best complement each other. Although you might have to ditch a great-sounding microphone because it does not work with any other, most tones can be achieved with two microphones only. The right tone could even be achieved with only one microphone, but it will be more difficult to obtain a stereo image of the guitar during mixing. If you cannot commit to a particular pair of microphones at the recording stage, the vision decided on during preproduction may not be clear enough. Always leave all the microphones plugged in during the recording, since different pairs may work better for different parts of the song, but always ensure that only two microphones are record-armed at any time. Limiting your recording to the best two microphones often means that they can be run through the best pair of preamps available. This setup is particularly useful in overdubbing studios, which often have only a couple of quality preamps.

Note that microphones can sound vastly different when placed off axis. For example, the difference in tone between one SM57 on axis and another off axis could even be so great that they end up being the best pair chosen.

8.4 Pulling the Right Sound

As with every live instrument to be recorded, getting the sound right at the source is crucial; therefore, the process to be followed is to do the following:

- Zero out all the knobs on the amp and guitar
- Choose the best pickup for the tone to be achieved
- Tweak the tone options on the guitar until you are satisfied with the result.
- Move on to changing the tone and EQ knobs on the amp until the best possible sound is achieved
- Choose the right microphone and experiment with positions
- Choose the right preamp/ADC combination

Since electric guitars are sometimes recorded with distortion, this process is very often altered, as the results can be unpredictable. If this is the case, rerun through these steps after finding the right distorted tone.

8.5 Running the Session

Most of the advice given in Chapter 7, "Bass Guitar," is relevant when you are recording an electric guitar player, that is, when possible, recording with the musician in the control room if the tone to be achieved does not involve interaction between the instrument and the amp (i.e., feedback). Tuning the guitar at the beginning of each song is also a good idea. Only the final parts should be kept in the session, so it is important not to record different playlists with this instrument. If there are timing problems, fix them right away. The only exception to this rule is for guitar solos. In this case, let the guitar player record multiple solos, but with the condition that he or she will have to pick or compose the final take during the same session. There is nothing worse than postponing that decision to the mixing stage, at which the engineer is left to decide on the compositional aspects of the song. If a guitar player cannot settle on a solo performance, he or she is probably not fully satisfied with any of the takes. This unhappiness could ultimately result in another recording session weeks later!

Depending on the vision, layering multiple takes of the same guitar part may be needed: a common overdubbing technique that is used to thicken up the guitars in the mix. Remember that everything comes at a price, and if the decision has been made to "quadruple-track" the distorted guitars in the chorus, this method will be at the expense of something else in the mix. For example, the drums may now sound a lot smaller, a result that could impact negatively on the energy of the song as a whole. In a rule of thumb, overdubbed parts sound thicker, but at the expense of "definition" and "punch." For an illustration of this concept (Figure 8.9; Audio 8.9 ▶), think of a snare drum that has been overdubbed four times. It will have a richer sustain part and will last a little longer, but all the initial transient's definition will be lost, giving a less than powerful sound overall.

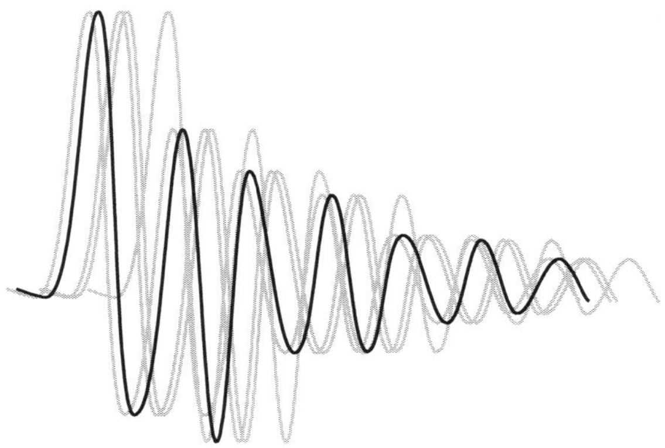

FIGURE 8.9
Overdubbed waveform.

The same concept applies to panning electric guitars. Panning one take to the left and the other to the right will achieve more separation and punch, while using both microphones of each take panned hard left and right will result in a thicker but less defined sound. See Chapter 13, "Panning," for more on this concept.

8.6 Creating Interest with Electric Guitars

8.6.1 Different Tones

In most productions, the guitars should be "moving" and creating interest. Whether it be consciously or unconsciously, musicians know about this concept and the engineer is often asked to record the verse with a different tone, guitar, or amplifier from that of the chorus.

Since all the microphones are set up in front of the cab, try noting two that sound very similar to each other and swap one for the other halfway through a verse. In this case, you will need to record with three microphones and break the "two microphones only" rule mentioned earlier.

The guitarist should also experiment with playing a particular part with a plectrum instead of his or her nails, or using different hand shapes to play the same chords.

8.6.2 Clean Version

Record a clean version of distorted parts by using a DI or double tracking with a clean tone. The second method usually works better, as it gives more control over the tone being recorded. Mixing this extra part underneath more distorted takes can work wonders in bringing back the dynamics and definition lost when you are using a lot of saturation. While clean recording is often done on acoustic guitars, using a clean electric guitar can often work. Beware that acoustic guitars are usually associated with pop or folk music and could therefore "soften" the feel of the song.

8.6.3 Micro-modulations

Subtle micro-modulations can work wonders on stale rhythmic parts. Flangers, phasers, filters, reverbs, and delays can be used creatively to add movement to those parts. These effects should be set to slowly evolve over the course of the song.

Chapter 9

Acoustic Guitar

While the musical genre and the vision will dictate the role of the acoustic guitar within the mix, knowing the techniques that can make it blend well or, on the contrary, cut through the mix, is essential. As always, remember that more time spent at the recording stage means less work during mixing.

9.1 The Room

The choice of room is very important for this instrument. Capturing the sound too "dead" or "live" can result in the guitar being hard to mix with the rest of the instruments. Although similar guidelines as those laid out in Chapter 6, "Drum Kit," apply when you are picking a room for acoustic guitar, it really comes down to the way the guitar sounds in the room. While the general guideline of "concrete sounds cold" and "wood sounds warm" often applies, it is important to always judge the sound of the guitar in the room in which it is being recorded. A nylon-string guitar may sound very warm when recorded in a small room with brick walls, and dull in a larger wooden room, for example. Use your ears to judge whether the room is appropriate.

9.2 The Instrument

All acoustic guitars have their own sound because of the different woods they are made of, their size, and their shape. For example, small-bodied guitars sound "smaller" than bigger ones. Another aspect of the instrument to take into consideration is the type of strings that are being used. Nylon strings tend to sound warmer than steel strings, but at the expense of presence and definition. The performer will usually bring his or her personal guitar, so the choice is often limited when it comes to this instrument.

9.3 The Microphones

Most stereo microphone techniques (described thereafter) will work well on this instrument. If the guitar has a DI output, it should be used systematically. With modern DAWs, it does not cost anything to record this extra track. While it rarely sounds great on its own, this DI output can prove to be very useful when drastic EQ curves such as adding a lot of body or presence must be applied during mixing. It is often easier to keep the original characteristics of the signal captured with microphones while adding what is missing with the DI track. Another reason to record the DI output is that it lends itself to amp emulations a lot more easily than signal recorded with microphones.

When it comes to microphones, using only a pair should be the preferred method. Just as with a guitar cab, most tones can be achieved with a pair of microphones. The three main stereo microphone techniques used for acoustic guitar are mid-side, XY, and spaced pair.

9.3.1 Mid-Side Pair

This is a great technique, as it requires the engineer to listen only to the center microphone when deciding on the recording position (Figure 9.1). If the center microphone on its own does not sound great, the overall result will not be much better. Because of this factor, spend time getting an assistant to move the center microphone while you are listening to it in the control room. The choice of polar patterns will also greatly affect the final sound:

- A cardioid center microphone, if placed close enough to the body of the guitar, is subject to proximity effect, which can be desirable if the guitar sounds too thin and needs extra low end. This pattern is also is the least subject to picking up the sound of the room, so it should be favored when you are recording in an unflattering room (Audio 9.1 ▶).
- An omnidirectional center microphone will be most subject to room sound, which can help achieve a great mono tone while still keeping ambience and clarity in the recording (Audio 9.2 ▶).
- A bidirectional center microphone is a good compromise between the previous two patterns, keeping the focused aspect of a cardioid and the open sound of an omnidirectional (Audio 9.3 ▶). Depending on the microphone model, the widest stereo image can sometimes be achieved by using this polar pattern.

A MS pair tends to blend better within busy mixes than do other microphones. It does not have an "intimate" quality, but rather one that makes it easy to fit the guitar as a background instrument when the mix is already cluttered. If a feeling of closeness or intimacy is sought (as is usually the case for acoustic singer–songwriter music), an XY pair should be used.

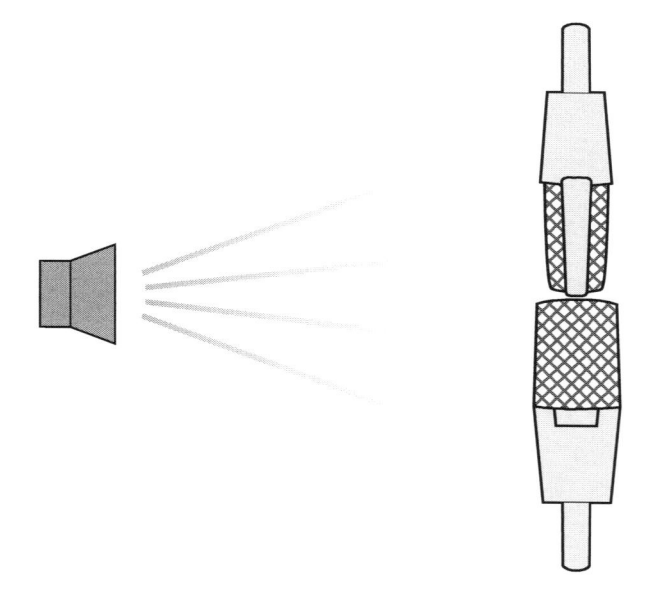

FIGURE 9.1
MS pair of microphones.

9.3.2 XY Pair

An XY pair of small-diaphragm microphones carefully placed can give great results (Figure 9.2; Audio 9.4). This technique also allows for great mono compatibility, but the stereo image tends to be narrower than when an MS pair is used. Placing the two microphones on a stereo bar is helpful for moving them easily without having to spend time in fine-tuning their relative positions. Most engineers will place an XY pair roughly where the guitar neck meets the body. This placement, which allows for a good balance between precise (string noises) and full (body resonance) tones, is only a starting point, though, as getting an assistant to move the microphones while you are listening in the control room is necessary for finding the best position. As mentioned earlier, this technique is the favorite when an intimate sound is required.

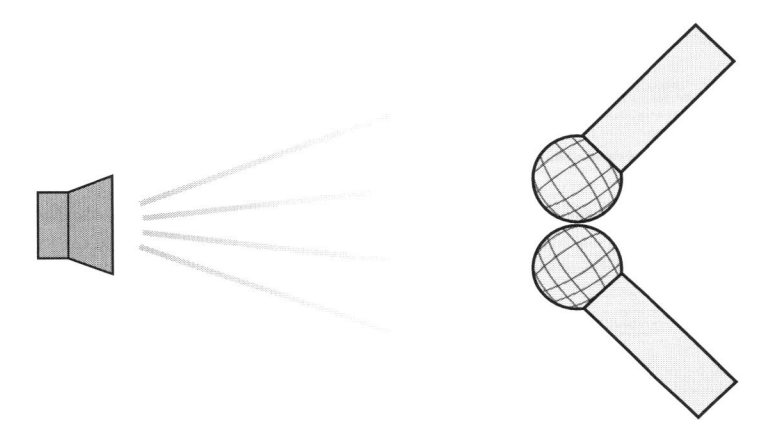

FIGURE 9.2

XY pair of microphones.

When you are working in a great-sounding room, swapping cardioid microphones with bidirectional ones (Blumlein technique) can also give very good results (Audio 9.5). The swap will add ambience and natural reverb to the sound, and the sound achieved will be closer to that of an MS pair in terms of how it can blend with other instruments.

9.3.3 Spaced Pair

Although this technique is often the last one tried, a spaced pair of microphones can sometimes be used to give great results (Figure 9.3). There are no real rules here, and different engineers work in different ways. Some of the common techniques include the following:

- Placing two cardioid or omnidirectional microphones next to the performer's ears to place the listener at the same listening position as the original performer. If cardioid polar patterns are used, this configuration is essentially an ORTF (Audio 9.6).

- Placing one microphone close to the guitar (a large-diaphragm cardioid microphone where the neck meets the body of the guitar is a good place to start), and one roughly a meter away to give a sense of naturalness to the instrument while keeping definition from the close microphone (Audio 9.7 ▶).
- Placing one large-diaphragm omnidirectional microphone where the neck meets the body of the guitar for treble, and one large-diaphragm cardioid microphone on the sound hole for bass (Audio 9.8 ▶).

This technique can also be modified to work in conjunction with an XY pair. If you are placing the pair so that no microphone is pointed at the hole, adding a "bass" microphone on the hole and panning it center can sometimes help get a fuller tone. This placement is particularly useful for mixes that have a very sparse arrangement and do not contain instruments filling the low end of the frequency spectrum. Always try to capture the required tone with two microphones, and add the third one only if necessary.

Be careful when you are using non-coincident microphones on a moving sound source such as an acoustic guitar. Because of the arrival-time difference of the sound between one microphone and the other, slight movements of the instrument itself can introduce uncontrollable phasing issues in the recording.

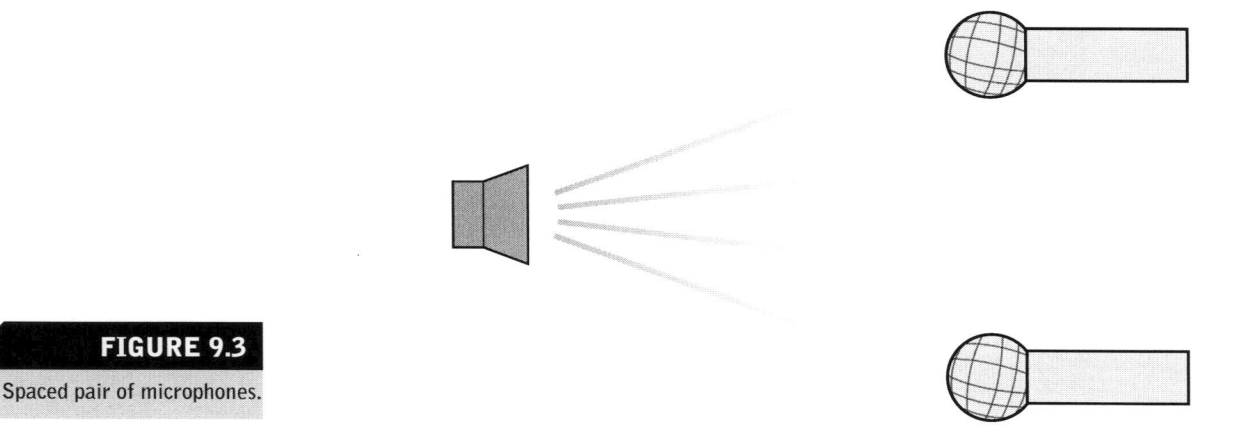

FIGURE 9.3
Spaced pair of microphones.

For extra tips regarding recording acoustic guitar and vocals simultaneously, see Box 9.1.

> **BOX 9.1**
>
> *It is sometimes necessary to record both acoustic guitar and vocals simultaneously as the performer may be accustomed to this way of playing from years of performance practice. If this is the case, achieving separation between the guitar and voice may be difficult. The best separation possible can be achieved through the use of bidirectional polar patterns with all microphones. Recording the guitar using Blumlein or MS (with bidirectional center microphone) techniques and the voice with a single bidirectional microphone will yield the most separation possible. You will need to ensure that the dead side of the guitar microphones are pointed directly at the mouth of the singer, and the dead side of the vocals microphone is pointed directly at the guitar.*

9.4 The Processors

Compression can be of great benefit to an overly dynamic acoustic guitar, but compressing this instrument can be very challenging due to the artifacts associated with dynamic range reduction. Using compressors that are too fast can result in unnatural pumping, while using soft clipping can distort the tone if the clipping is not carefully applied. Optical leveling amplifiers are very useful for this instrument. Because of the way the attack and release speeds change with the input level, these processors act very smoothly on the dynamics (Audio 9.9 ▶). If more compression is needed, following this compressor with a more aggressive one or limiter can further help with dynamic range reduction.

When you are trying to fit acoustic guitars within busy mixes, it can be beneficial to use an MS EQ to high-pass-filter the center channel higher than the sides. This step effectively creates more space in the center of the mix for kick, bass, and any other bass-heavy instrument. It also creates the illusion of a wider acoustic guitar. As a general rule, leaving an extra octave in the sides channel works best in most cases (Audio 9.10 ▶).

The sound of a plectrum or fingernails strumming an acoustic guitar can be lowered without affecting the fundamental tone of the playing by using split-band de-essers or multiband compressor. By triggering and compressing only the top end of the guitar, you can reduce the volume of the high frequencies with little effect on the tone of the guitar. This method can be very effective in making the acoustic guitar less aggressive (Audio 9.11 ▶).

9.5 Creating Interest with Acoustic Guitars

Acoustic guitars have the potential to create subtle movement in the mix. Useful techniques for the engineer or guitarist include the following:

- Swapping guitars for different parts of the song to give a different tone
- Swapping pick thickness or type of strumming
- Swapping the strings on the acoustic guitar with the thinner set of strings of a twelve-string guitar for a higher/thinner tone (Nashville tuning)
- Getting the performer to play while standing rather than sitting
- Using EQ and stereo-width automation, which work well because they allow the guitar to be more present or, on the contrary, leave more space for others in busier parts of songs
- Playing the same part at a different place on the neck (Figure 9.4), or tuning the instrument differently to force different hand shapes to give a different feel to the same performance.

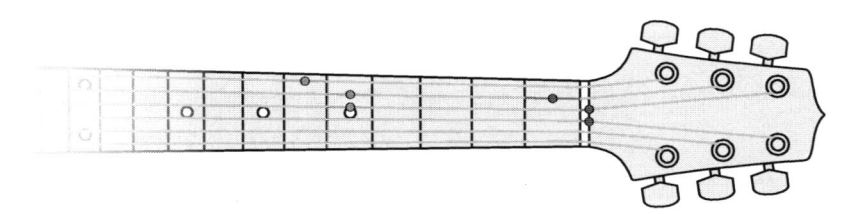

FIGURE 9.4
Different finger positions for the same chord.

Chapter **10**

Vocals

In most genres of music, vocals are the most important element of the song. If recorded right, they can be easy to mix, but a poor recording often results in hours of work of trying to make them fit within the mix. Vocal engineering (recording and mixing) is a job in itself, and there are engineers who specialize in this task. This call for expertise does not mean that it is impossible to get the same quality as the big guns in the industry, but rather that they require a lot more work than other instruments in order to sit properly in the mix.

10.1 The Room

Recording vocals in a booth or a very dead acoustic environment allows for most control at the mixing stage. With good-quality digital reverb units now readily available, recording in a particular room is less of an issue, as the band's virtual acoustic space can be fairly accurately replicated at a later stage (more on this topic in Chapter 16, "Reverberation and Delays"). The exception is when recording a vocal ensemble, in which case the choice of room is very important. A large choir recorded in a church will sound very different from one recorded in a highly-damped room. Finally, for single-singer recordings, the placement of the microphone within the recording room needs to be determined by recording short parts at various points in the room and noting where the voice sounds best. There are no rules with this except that each test recording must be made with the singer at the same distance from the microphone to remove this variable when you are judging sound quality.

10.2 The Instrument

While there is not much room for modifying the tonal aspects at the source, some guidelines still apply when recording vocals. The first step is to ensure that the singer warms up before the session. This stage is especially important when you are recording in the morning, as the singer has fully rested vocal folds. Singing in the appropriate range for a singer is also very important. Although this range should have been decided during preproduction, tuning the whole song up or down a few semitones can work wonders in obtaining the best tone from a singer. These adjustments can be used to ease the performance or, on the contrary, to push the singer out of his or her comfort zone to enhance the message of the song.

Singers can sometimes achieve more powerful tones when their heels are raised, a position that can enhance posture and help push air out more easily. For female singers, asking them to bring high-heeled shoes to the session should not be an issue (wedge-type shoes should be favored, as stilettos can cause stability issues) (Figure 10.1; Audio 10.1 ▶). For male singers, using a pack of A4 printing paper can also work.

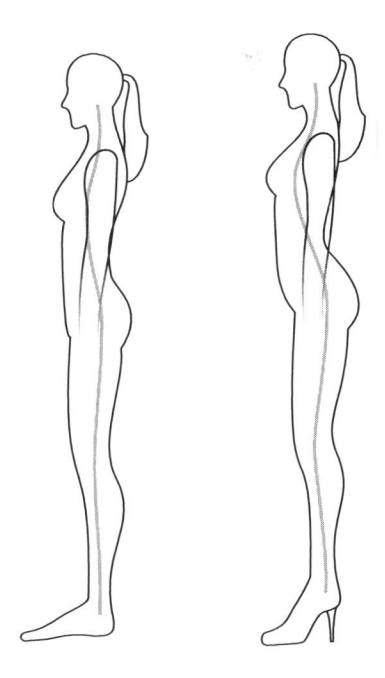

FIGURE 10.1

Effect of wearing high heels on posture.

To ensure the singer always sounds as good as possible, leave a tall glass of water in the recording room. Do not ask if the singer wants water, as singers often say no. Instead, leaving the glass in the recording room allows the singer to drink without having to stop the session. Water lubricates the throat and helps achieve more consistent performances.

10.3 The Microphones

While large-diaphragm condenser microphones are the most common type for recording vocals, trying different models when recording works better than any amount of equalizing in making the vocals fit within the mix. For example, a lower-quality small-diaphragm microphone may suit the voice better for a particular song.

While it is always recommended that the singer knows the part when he or she starts recording, some performers like to have a lyrics sheet hung up next to the microphone (Box 10.1). Each singer has a "better side," where the voice projects more strongly and sounds clearer (Audio 10.2 ▶). The lyrics sheet should be hung up on the opposite side so that the singer's "good side" is facing the microphone if he or she looks at the lyrics while recording. The easiest way to find out which side sounds better is to ask the singer to sing a constant note while you listen to both sides from about thirty centimeters away. One side always has more power than the other.

AUDIO PRODUCTION PRINCIPLES

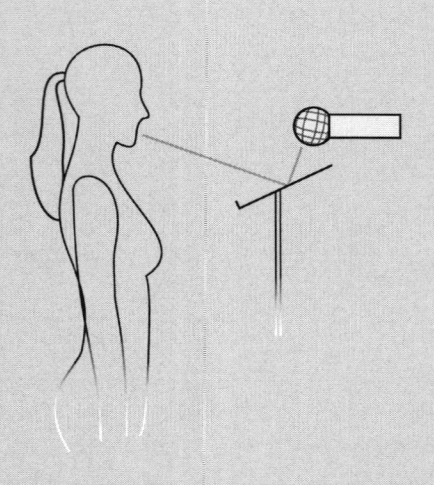

BOX 10.1

Print lyrics in as large a font as possible with double line spacing, and leave a pencil and eraser in the recording room. This setup will allow the singer to write vocal inflections or notes which may help him or her deliver a better performance. Furthermore, ensure that the lyric stand is not angled towards the microphone as the reflections from the voice could create comb filtering (Figure B10.1).

FIGURE B10.1
Voice reflected on lyric sheet stand.

As previously mentioned in Chapter 4, "Microphone Basics," different polar patterns will affect the recorded vocal tone and should be used accordingly. Cardioid and bidirectional polar patterns can help give a more intimate feel to the performance when proximity effect is used. They often give the impression that the singer is closer to the listener and can add body to the voice. Be careful when using these polar patterns with singers who are seasoned live performers. Singers with good microphone technique will often pull away from the microphone as they sing more loudly, and that movement can make the recorded tone inconsistent. If a cardioid or bidirectional polar pattern is used, ask the singer not to move away from the microphone and ensure that the preamp gain is low enough not to distort on loud notes. An omnidirectional pattern will have a more open sound and can be used in various ways. Singers with excessive bass in their voices can use this pattern to sing close to the microphone without being subject to proximity effect. This polar pattern is the usual one in recordings in which a more open sound is required. Backing vocals are also often recorded by using this polar pattern. Different placements of the microphone in relation to the mouth can also change the way the vocals will sit in the mix (Audio 10.3 ▶). Recording close to the microphone will give a more intimate and "closed" sound, while recording farther away will give a more "open" sound. A good starting point is two fists between the microphone and the singer.

Placing the microphone slightly to the side and angled toward the mouth will reduce sibilance, because of the directionality of high frequencies. Regardless of the polar pattern used, the off-axis position may also result in the overall tone being less natural (remember that even omnidirectional microphones can be colored off axis). If placing the microphone to one side, make sure it is the singer's good side, as previously identified for lyric-sheet placement (Audio 10.4 ▶).

Placing the microphone lower than the mouth (angled up) tends to give a brighter tone to the voice, somewhat equivalent to a presence boost on an EQ, while placing the microphone higher than the mouth (angled down) can help tame an overly dynamic performance, working similarly to a soft compressor (Audio 10.5 ▶). If placing the microphone lower than the mouth, ensure that the singer does not angle his or her head down, as this position will cut the throat's airflow and will result in a less powerful performance.

When you are using large-diaphragm condenser microphones, it could be useful to mount it upside down. There are many reasons for doing so. The first is that it may help reduce pops (the loud bursts of air that happen when singers are pronouncing the letters "p" and "b"), which are directed slightly lower than the mouth. If the microphone's body is in line with the singer's eyes, it may also help with ensuring, as previously discussed, that the singer does not angle his or her head down and cut the airflow. If a lyric sheet is used, there is also more space to place it correctly below the microphone. Finally, if tube microphones are used, the heat generated will rise and not affect the capsule's temperature. Note that if the artist sings very close to the microphone, this heat could be used to remove the moisture generated from breaths, which can sometimes affect the microphone's performance.

Do not forget to use a pop filter (Box 10.2) to ensure that the pops are attenuated. It is often preferable to mount the pop filter on a separate boom stand to ensure that the microphone does not shake if the singer touches the pop filter. There are two types of pop filters, which sound slightly different from each other: mesh and metal grill. Mesh filters work by slowing down the velocity of air (and to some extent sound) and can sometimes slightly darken the voice, while grill filters work by redirecting the air downward, keeping the tone intact. Metal grill filters need slightly more distance from the microphone to effectively remove pops, while mesh filters can be positioned closer if an intimate sound is required.

BOX 10.2

If the studio does not have a pop filter, placing a pencil in front of the microphone's capsule can help break the bursts of air going to the capsule (Figure B10.2; Audio B10.2 ▶). This technique can extend to the recording of other instruments that push a lot of air at the sound source, such as a kick drum.

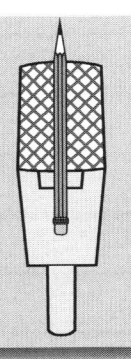

FIGURE B10.2

Do-it-yourself pop filter using a pencil.

10.4 The Processors

If recorded right, vocals can be mixed with little processing. Recording with a high-quality microphone and preamp does not necessarily give a better sound on its own, but it often allows the engineer to equalize and compress a lot more easily. Although each recording is going to require different processing, the following mixing tools are often used on vocals (in order of processing):

- Light volume automation: This technique is used to even out full words that are too loud (or not loud enough). It is common practice to ensure that most words peak around the same level and that breaths are reduced in volume (this last step is not always necessary). This process is done to ensure that compressors inserted later in the vocal chain are not being pushed too hard.
- Fixing EQ: Use high-pass filtering to get rid of sub-bass frequencies and low-pass filtering to make the mids stand out a little more. If needed, equalize out specific frequencies to get rid of nasal or harsh qualities. If you are working with a DAW, stock plug-ins and surgical EQs work well for this task.
- De-esser: Sibilance should be dealt with during recording by placing the microphone off axis and slightly to the side. If a de-esser is needed after the recording is done, it should be inserted at this stage. Beware of overusing this tool, as it can give the singer a lisp. If necessary, use a second de-esser later in the chain (using a different brand of de-esser works best if multiple instances are required).
- Limiter/field-effect-transistor (FET) compressor: These tools are used to even out the performance and get rid of loud syllables within words. It is the next step in dynamics control after volume automation. The limiter should barely be active, and the gain-reduction meter should move only with the loudest parts of words. Common hardware compressors used for this task are the Urei 1176 and API 525.
- Optical compressor: Opto compressors are used for overall smooth compression. If the volume automation and limiter have been done properly, this compressor should be constantly achieving a few decibels of gain reduction. If the opto compressor reads 1–3 dB of gain reduction at all times when the voice is present, then the previous dynamics processors were set well for achieving a very steady vocal in the mix. Famous units used for this process include the Teletronix LA2A and TUBE-TECH CL1B.

- De-esser: If the added compression has brought back some of the "s" sounds, use a second de-esser here, or edit them out with volume automation.
- EQ: Using high-quality equalizers are necessary to put the best qualities of the voice forward. Neve, SSL, Pultec, and other smooth-sounding units should be used here to add an "expensive" sonic print to the recording.
- Multiband compressor plus EQ: Compressing the "air" present in the voice (5–10 kHz and up) can give a more polished sound to the recording. Bring the frequencies that have been compressed back with a high-quality EQ rather than raising the gain of the compressed band to imprint more of that expensive sound. Pultec-style EQs work well for this.

10.5 Running the Session

Since vocals are the most important part of most songs and will need a lot of treatment, it is a good idea to record and edit them in a DAW session separate from the rest of the song. Exporting the instrumental version of the song and reimporting it into a new session will free up computing resources and ensure that the system is very responsive for the amount of work to be done. This stage will require the song to be balanced in a way that the singer is comfortable with. Most singers require a basic mix so that all instruments can be heard. When recording different takes, keep in mind what the vision of the song is and how many doubles, unison, and harmonies of each phrase are needed. If these factors are not planned ahead, too many takes may be recorded and the workload will ultimately be increased. Since the vocals are recorded into a new session, recording different parts onto different tracks can help with seeing where the session is headed. Creating subgroups at the recording stage for parts such as "verse doubles," "chorus low harmonies," and so on, helps in ensuring that editing time is cut down as much as possible.

Recording singers requires as many psychology skills at it does engineering (for an example, see Box 10.3). Because the goal is to get the best performance from the singer, a lot of positive feedback must be used to make the performers feel good about themselves.

> **BOX 10.3**
>
> *In order to get a vocal performance that reflects the lyrical content, ask the singer why he or she has written the song and what the meaning behind the lyrics are. By talking through it, the singer may remember the feeling he or she had when the song was written, a memory that can help with delivering an appropriate performance.*

Giving singers access to the volume of their own headphones is a good start in making each singer feel comfortable. Any cheap headphone amp can do the job as long as it is left in the recording room for the singers to access. Giving them access to the light dimmer, if available, can also help. Depending on their personalities, some will turn the lights down (or completely off), which can sometimes be an indication of how open to comments they will be. Dimming the lights could be indicative of an insecurity in their ability to perform, which in turn means that the engineer should be extra careful when giving feedback about their performance. It is sometimes also necessary to dim the lights in the control room if the engineer is not alone. More people in the control room mean more discussions, and possibly laughing. It can be extremely distracting for a singer to see people talking and laughing in the control room, not knowing whether the discussion topic is the performance.

Some singers can be very critical of their performance, which can stem from their own insecurity or lack of experience in the recording process. If this is the case, you will need to find ways of making them very comfortable during the recording process. The recording environment should be inviting (through using the right lighting, room temperature, and so on), and their attention should be steered away from the performance. When talking to them, steer their attention towards elements such as how carefully placed the microphone is for what you are trying to achieve or how well the preamplifier sounds on their voice. Once you have recorded a couple of takes, make them listen to those. Make sure you are smiling a lot before they enter the control room and during the listening, tell them the takes were really good, and how much fun you are going to have processing their voice to make it sound even better during mixing. Overall, this step can give singers extra confidence in their ability, and in you as their engineer.

Singers often want to record with reverb on their voices. Avoid this technique if you can. Adding reverb during the recording often means that the singers are relying on the effect to hide tuning issues, and they may modify their performances to suit the chosen reverb. The only exception is when the reverb has been carefully chosen to be as close as possible to the final reverb that will be used in the mix, and balanced properly. In this case, reverb can actually help singers deliver a performance appropriate for the piece. If reverb is being used during recording, it is important to monitor the vocals dry to ensure that the performance can be judged as accurately as possible by the producer.

If the performance is sharp or flat, a change in vocal tone can often fix the issue. Instead of asking singers to sing in tune (which will not help with the singer's confidence), ask to change the tone of their voice. If the singing is sharp, ask to put less power in the voice or sing in a darker tone. Lowering the volume of the music in the headphones can also "force" singers to sing more softly, as they will need to focus more on what they are hearing. For flat performances, ask them to put more power into their voice or sing in a brighter tone.

Similarly, raising the headphones volume of the music can also force singers to push their voice in order to compete with the song's volume. Another trick that can be used is to ask singers to wear only one side of the headphones, allowing them to tune their voice to what they are hearing from the room rather than the headphone return only.

In some cases, singers will insist on recording without headphones, having the music being played through loudspeakers instead. While this approach is not ideal in order to obtain the cleanest recording possible, you should never refuse this request, as the performers still need to feel as comfortable as possible. There are two main techniques that can be used when recording with a loudspeaker. The first technique requires only one speaker aimed directly at the "dead" side of the microphone's polar pattern (back for cardioid, side for bidirectional). Once the vocals have been recorded, record another take of the song playing through the loudspeaker without singing (keep the performer in front of the microphone, though). In the DAW, reversing the polarity of this "music only" track should cancel out most of the bleed on the recorded vocal tracks. The main issue associated with this technique is that no dynamics processing such as compression can be done in the record path, as the music bleed must remain the same in the voice and polarity-reversed bleed tracks. Note that if any editing or processing needs to happen post-recording, the polarity-reversed bleed track and each vocal take must be printed together onto a new track for this technique to work. The second technique requires two loudspeakers playing the music with the polarity reversed in one. Placing the microphone halfway between the speakers should yield very little bleed as one speaker is 180 degrees out of phase with the other. For this technique to work well, the microphone may need to be moved by the assistant engineer until the best position (the position yielding the most music cancellation) is achieved. Unlike the first technique, dynamics processors can be placed on the record path of the vocal microphone.

Unless there is real momentum with the recording of a particular song, give the performers a short break between parts of the song. This allows them to drink water and rest their voice. Longer breaks between songs are often necessary, too, to also give yourself time to start sorting through takes and ensure that everything needed for a particular song has been recorded.

Songs may be recorded in different ways. In a rule of thumb, get the singer to record the whole way though once or twice before advising on a particular method. More often than not, recording the song part by part is preferable. Starting with the first verse, then moving on to the second, and so on, can help ensure that the singer is concentrating on obtaining the best performance for each part. Choruses are often recorded one after the other, as there are so many similarities between them. Try to avoid recording harmonies in this first recording phase if a lot of layering needs to be done, as it can make singers feel that the recording is dragging and can bring their morale down. Instead, recording the main voice and doubles for each part will keep up a certain momentum.

A good habit to get into is to color-code the recordings on the spot, so you can greatly reduce time spent later when compiling all the different takes, as they are already sorted. Working with three different colors (average, good, great) is often enough and helps you visualize when enough takes have been recorded. Some DAWs also allow to rate clips as they are being recorded, a method that can also work.

At times, the singer will struggle with a particular phrase. While it is possible to record a phrase word by word, try every other option before working this way (give the singer a break, record at a slower tempo, etc.). It will work in getting the words down, but will never capture the "feel" of the performance.

One last word of advice for when you are running a vocal recording session: always be in record mode! Singers often come up with great lines when trying out ideas and forget them the next second (as happens more often than you would think). If recording with Pro Tools, you can easily do so by turning on "quick punch."

10.6 Editing

Editing vocals takes can be a very long and tedious process. At the very least, it involves listening to the different takes and selecting the best lines for lead and backing vocals (if you have not previously done so with color coding). Depending on the genre, time-aligning words and removing sibilance from the backing vocals may also be needed. While there are plug-ins that can achieve this edit automatically, the artifacts added when you are using these tools are often too obvious. With a bit of practice, this task should not take more than a couple of hours per song. Time-aligning vocals involves cutting words and syllables so they all start and end at the same time. This edit has the effect of strengthening the vocal performance and gives the illusion of one powerful voice rather than having the listener recognize that the singer has overdubbed him or herself. If possible, try to time-align words by using "cut/paste/crossfade" rather than time stretching. The overall result is usually more natural sounding. While you are editing backing vocals to follow the lead, it is also preferable to manually de-ess them: cutting all "s," "f," "x," "z," and "shh" sounds. Since all words and syllables are time aligned, sibilance will add up and can sound quite harsh. Because the lead vocal still contains sibilance, the intelligibility of the voice will be given by the lead rather than the backing vocals. The overdubs should also be de-breathed and the letters "t," "k," "d," and "p" should be removed. Unless these letters and breaths are perfectly time aligned, their doubling up can make the performance sound sloppy. With a bit of practice, you will be able to visually recognize all these sounds (the waveform of consonants is denser, as they are mostly made up of high frequencies). If the backing vocals sound a little odd when played back with the lead, fading in the previously removed "s" sounds can bring back some naturalness without adding back too much sibilance.

When you are editing syllables together, there are a few rules to follow in order to get the best result. The first is that "s," "shh," and other sibilant sounds can be cross faded anywhere without many artifacts, so shorten or elongate these sibilants at will (without using time stretching). As long as the crossing point has equally loud parts on either side, a simple cross-fade will work (Figure 10.2).

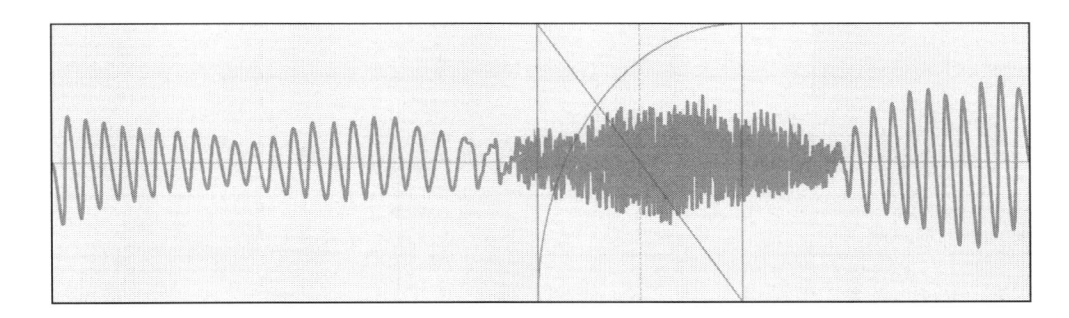

FIGURE 10.2
Cross-fading sibilant sounds.

Another useful trick to know is that the start of a new syllable can also be fairly "grossly" cross faded to the end of the previous syllable without much impact on the naturalness of the sound (Figure 10.3). This technique means that if a syllable needs to be shortened, dragging the start of the next one across and cross-fading will work best. The transition from one syllable to the other will mask the cross-fade.

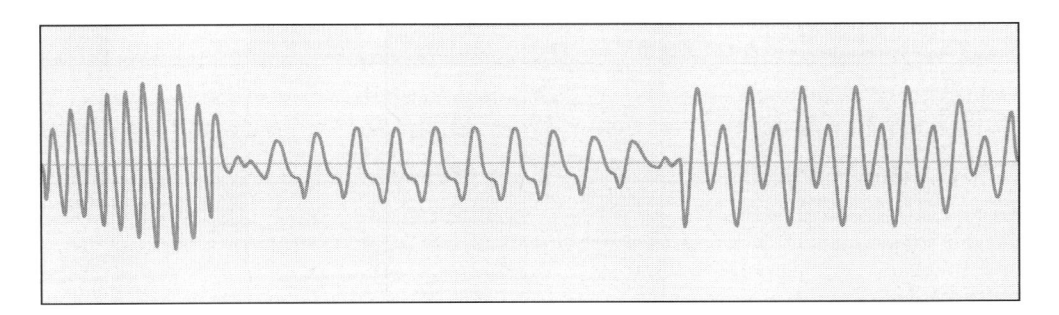

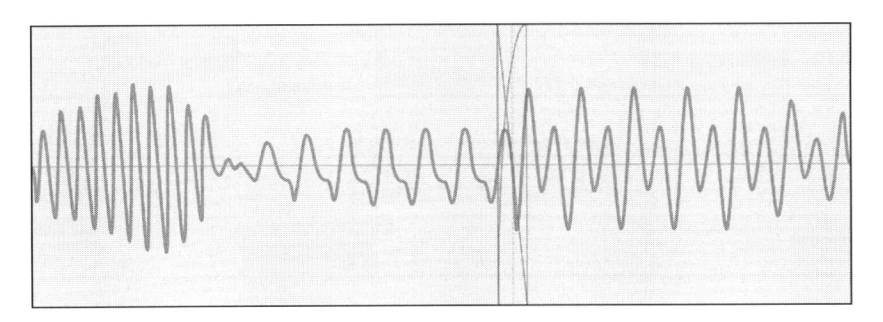

FIGURE 10.3
Shortening syllables.

The hard parts to shorten or elongate are the middles of vowels. In order to do this, find a recognizable pattern in the waveform and move the clip so that the pattern remains constant after the move (Figure 10.4). Always cross-fade these sounds at the zero-crossing point, starting with a short fade and making it longer if necessary. When a vowel needs to be stretched too far, this technique may introduce unwanted sonic artifacts, in which case time stretching can help.

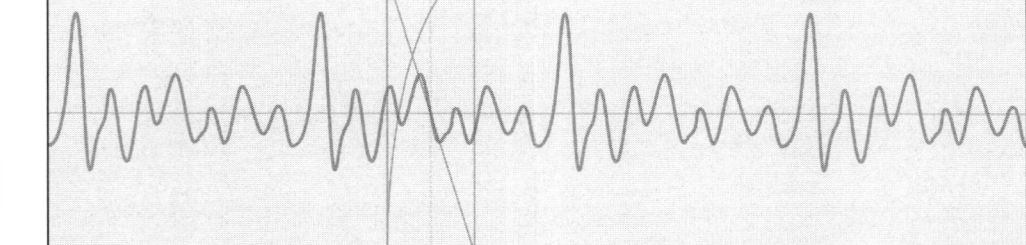

FIGURE 10.4
Elongating vowels.

10.7 Tuning

It is undeniable that most modern music uses some kind of vocal tuning processors. While the use of such processors is often controversial, and a lot of engineers consider it as "cheating," note that this use is no more unnatural than editing drums in time, or double-tracking guitars. The main issue to consider is whether a perfect performance is appropriate for the genre, as some music such as grunge can actually benefit from imperfections. Nonetheless, if a syllable seems too far out of tune, it is the engineer's job to fix it.

Some processors such as Melodyne even go one step further by offering a full vocal production tool rather than simply one to use for tuning only. Using these processors can be compared to having a vocal coach in the studio during the recording. They allow the producer to change the length of syllables, the pitch curve between syllables, the inflection at the end of certain words, and the amount of vibrato in the voice. While this is more of a production tool than an engineering one, using it can help mold the performance to suit the song perfectly. For example, if the last word of a phrase goes up in pitch toward the end but the lyrics are portraying a sad feeling, it may be more beneficial to change this vocal inflection to go down so the performance follows meaning.

10.8 Mixing Tips

10.8.1 Panning

Pan high harmonies toward center and low harmonies further to the sides (Audio 10.6 ▶). This step gives a wider stereo spread to the vocals, since everything else in the mix works in the opposite fashion (bass in the center and treble toward the edges).

10.8.2 Microphone Choice

The choice of microphone when recording vocals will greatly affect the amount of EQ that can be applied at the mixing stage. Some microphones will allow for far less equalizing than others: thus expensive microphones are often used for vocals. Great-sounding results can be achieved with an SM57, but it will not lend itself very well to EQ moves during the mixing stage.

10.8.3 Vocal Equalizing

The human voice contains mostly 1 kHz–2 kHz information, but boosting these frequencies on the vocals often gives unnatural results. Instead, try to equalize those other instruments in the same range that could be masking the voice (such as guitar, piano, and synths) while listening to the effect they have on the vocals. You will almost always find that cutting some of those frequencies in midrange-heavy instruments is more effective than using EQ boosts on the vocal track. In order to get a more present vocal sound, you may not need to cut 2 kHz on the piano, but 800 Hz instead. Different voices behave differently when surrounded by instruments and equalized, so make sure to treat each new voice differently. In any case, and since we (humans) are very "tuned" to the sound of the voice, minimal processing on this instrument should be used.

10.8.4 Top-End Boost

Vocals often allow for high amounts of top end to be added with EQs. Using high-quality EQs such as SSL, Neve, or Pultec units allow for a lot of high frequencies to be added without making the vocals too harsh (Audio 10.7 ⓘ). It is not uncommon to cut high frequencies with stock DAW EQs, then add them back with high-end EQs to impart their sonic characteristics to the voice.

10.8.5 De-Esser

While wide-band de-essers often work best for vocals, if heavy compression is necessary, using a split-band de-esser with the frequency to be reduced set higher than the detection frequency can work wonders. This process ensures that only the highest frequencies are reduced and can sometimes allow for smoother de-essing than if both the detection and reduction frequencies are set around the usual 7 kHz.

10.8.6 Parallel Compression

Parallel/upward compression can be used on vocals to emphasize detail, mouth noises, and breaths, for example (Audio 10.8 ⓘ). Unlike using parallel compression on drums, in which the aim is to mix the unprocessed transient with a highly compressed sustain, this way of processing voice will raise the low-level information such as mouth noises and blend those with the more tonal elements of the voice.

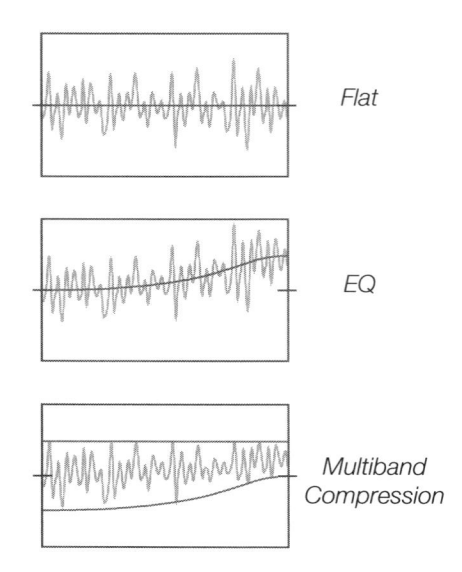

10.8.7 Multiband Compression

In order to hear the "air" quality of the voice more evenly throughout the performance, use a multiband compressor set to work heavily on high frequencies (Figure 10.5; Audio 10.9 ▶). The aim is to be constantly compressing the top end (5 kHz and up). Using low ratios, fast attack, slow release, and a healthy amount of gain reduction means that the very top end of the voice will be evened out and therefore the "air" will be heard at all times.

FIGURE 10.5
Equalizing versus multiband compression to enhance high frequencies.

10.8.8 Whispered Double Track

In order to ensure that the top-end sizzle of the voice is constantly present, you can use the technique of double-tracking the vocals with whispers. These tracks can then be highly compressed to achieve a similar effect to the one previously listed by using multiband compression. An extension of this technique involves sending the whispered double to reverb instead of the main voice, as a way to obtain a different reverb quality.

10.8.9 Side-Chain Compress Reverb Tails and Delays

Long reverb tails and delays on vocals can add mud to the performance. These effects can still be heard while new lyrics are sung. The most common technique used to clean up vocals after adding such effects is to use a compressor set to work on the FX bus with its side-chain detection triggered by the dry vocal (Audio 10.10 ▶). Doing this cleanup means that while the vocals are being heard dry, no FX is clouding the performance, and only when the voice has finished singing will the tail of those FX come out.

A similar result can also be achieved by using a split-band de-esser or multiband compressor on the reverb return and setting its compression frequency in the main vocal range (1–2 kHz). Using this technique means that only the low mids and the top end of the reverb will be heard. If this range affects the reverb too much when the voice is not present, using both techniques mentioned in parallel will achieve the best results, keeping only high frequencies when the vocals are present, then the full reverb when they are not.

10.8.10 Slapback Delay

Modern vocals often benefit from a light slap delay between 60 ms and 120 ms as a way to add dimension and sheen without blurring the signal (Audio 10.11 ▶).

To add dimension, you can often make the delay a little dull by using low-pass filters and vintage saturation. To add sheen, you can widen the delay by using the Haas effect and other stereo-widening effects, further processed with short reverberation, saturation, and EQ. If you are adding EQ to the delayed sound, boosting 1 kHz to 2 kHz can give a modern edge to the vocals without making it sound unnatural.

10.8.11 Heavy Distortion

A common effect used in order to give more presence and attitude to a vocal take is to add distortion, bit crushing, and other harmonic enhancement processors. If heavy processing is applied, there can sometimes be harsh distorted tails added to words. If you get this unwanted effect, running this process in parallel and using a gate before the distortion will help ensure that the processing is heard only when the voice is present and cut when there is no singing (Audio 10.12).

Chapter 11

Synthesis Basics

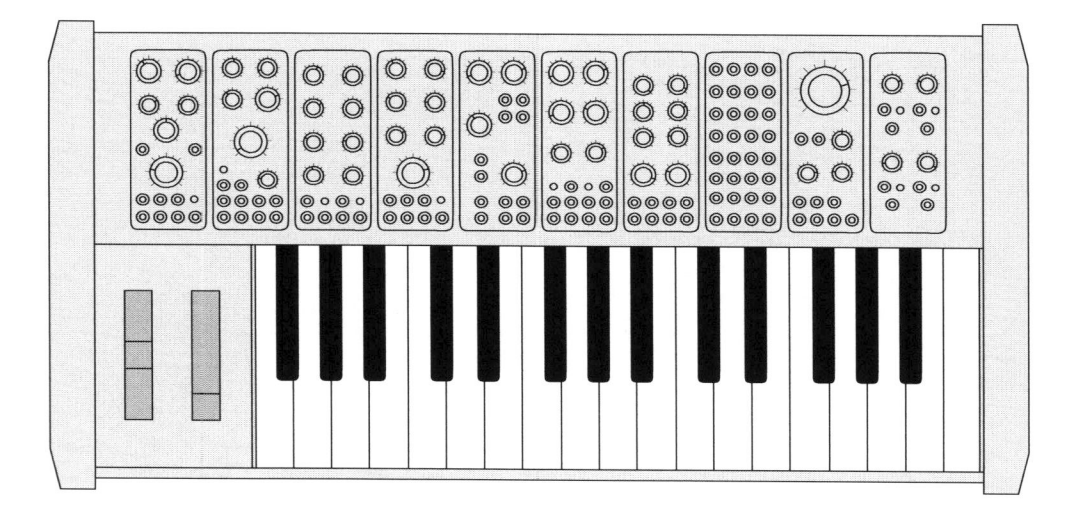

Virtual instruments are a vital part of modern music production. Without a basic understanding of synthesis, sampling, MIDI, and other technologies associated with this topic, an engineer may find this gap in knowledge to be a vital limitation to his or her arsenal. To that end, this chapter aims to demystify the various components of common synthesizers and give examples of their uses in production.

In acoustic instruments, a physical element such as a guitar string or drum skin oscillates to create sound. In a synthesizer, the oscillator generates sound that is then further manipulated with effects. As with acoustic instruments, sound design using synths starts with the waveform created. In much the same way that different guitar models have different sounds, different synth models have slightly different basic waveforms and filter designs that give them their own unique sound. As an example, a Moog sawtooth wave looks somewhat rounded, whereas a Waldorf has a small spike at the start of each cycle. While a deep knowledge of the sounds associated with brands and models is not essential, an understanding of the common modules available in synthesizers is crucial in order to create the intended sounds.

11.1 Waveforms

Synthesizers generate sounds by using oscillators. The main waveforms used in synthesizers are the following:

11.1.1 Sine

This is the simplest waveform. It contains only the fundamental harmonic and is usually referred to as having a "smooth" tonal quality (Figure 11.1; Audio 11.1 ▶). When used an octave below a lead line, it can add weight without changing the tone of the sound.

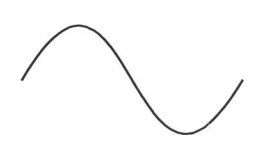

FIGURE 11.1

Sine-wave shape and frequency content.

11.1.2 Triangle

This waveform contains the fundamental and odd harmonics that quickly decrease in volume in the higher range of the frequency spectrum (Figure 11.2; Audio 11.2 ▶). The resulting sound is often called "woody" or "hollow" in comparison to the more complex waveforms.

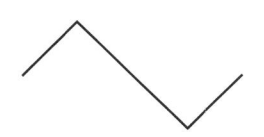

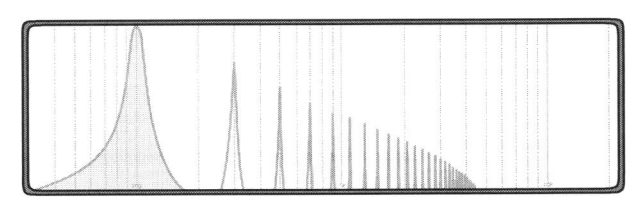

FIGURE 11.2

Triangle-wave shape and frequency content.

11.1.3 Square (and Other Pulse Width)

This waveform also contains the fundamental and odd harmonics, but at a higher level than in a triangle waveform (Figure 11.3; Audio 11.3 ▶). The sound of a square wave is quite distinctive, as it was used a lot in older 8-bit video games. While similar to a triangle in tone, it is a little more aggressive and "lo-fi" sounding. Varying the width of the pulse (the negative versus positive cycle lengths) changes the volume relationship of the different harmonics.

FIGURE 11.3
Square-wave shape and frequency content.

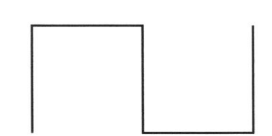

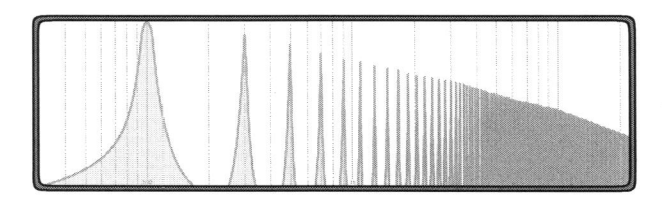

11.1.4 Sawtooth (and Super Saw)

This waveform is the most complex, as it contains both even and odd harmonics. It sounds brighter and more piercing than any other waveform (Figure 11.4; Audio 11.4 ▶). Sawtooth waves also form the basis of "super saws," which are slightly detuned doubles stacked together. The first super saw had seven oscillators detuned to give that very recognizable "trance pad" sound.

FIGURE 11.4
Sawtooth-wave shape and frequency content.

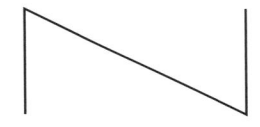

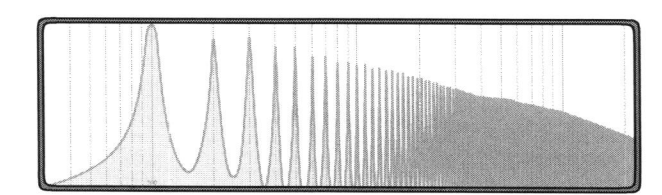

11.1.5 Noise

While not technically a waveform, noise is often used in sound design to add brightness without overly affecting tonal quality (Figure 11.5; Audio 11.5 ▶). When used in conjunction with a super saw, it can add thickness without having to use a lot of extra oscillators.

FIGURE 11.5
Noise example shape and frequency content.

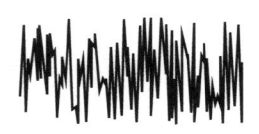

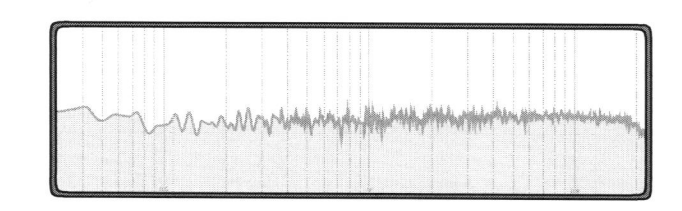

11.2 Filters

In subtractive synthesis, filters are used to remove frequency content from complex sounds (a process that is the direct opposite of additive synthesis, which adds sine waves together to form more complex waveforms). Each filter has a sound of its own due to a slight phase shift that happens around the cutoff frequency. Just as different EQs sound different, synthesizer filters add their own sonic quality onto the sound that is being run through them; a Moog filter sounds different to a Sherman filter for example. The slope or steepness of the filter curve is defined in "decibels per octave" or "poles" (1 pole = 6 dB/Oct). The steeper the slope, the more the effect and sound of the filter will be heard.

The most commonly found filters in a synthesizer are the following:

11.2.1 Low-Pass Filter

This filter cuts high frequencies and lets only the low-end pass through (Figure 11.6; Audio 11.6 ▶). The low-pass filter (LPF) can be used to turn a complex waveform into a sine wave.

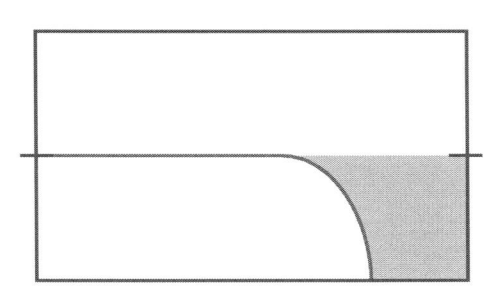

FIGURE 11.6
Low-pass-filter EQ curve.

11.2.2 High-Pass Filter

This filter cuts low frequencies and lets only the high-end pass through (Figure 11.7; Audio 11.7 ▶). The high-pass filter (HPF) can be used to "thin out" a complex waveform by removing its fundamental frequency.

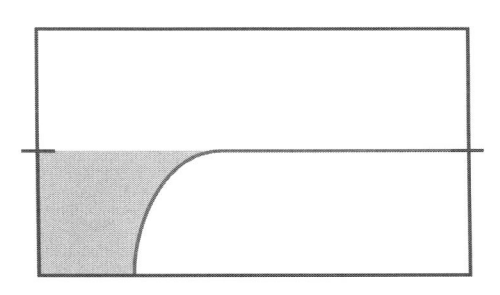

FIGURE 11.7
High-pass-filter EQ curve.

11.2.3 Band-Pass Filter

This filter is a combination of both LPF and HPF, where the LPF center frequency is set higher than the HPF center frequency (Figure 11.8; Audio 11.8 ▶). This filtering effectively removes both low and high end.

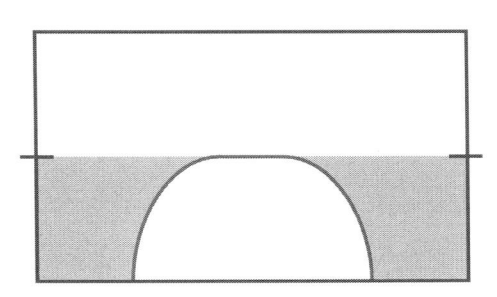

FIGURE 11.8
Bandpass-filter EQ curve.

11.2.4 Band-Reject Filter/ Notch Filter

This filter is a combination of both LPF and HPF run in parallel, where the LPF center frequency is set lower than the HPF center frequency (Figure 11.9;

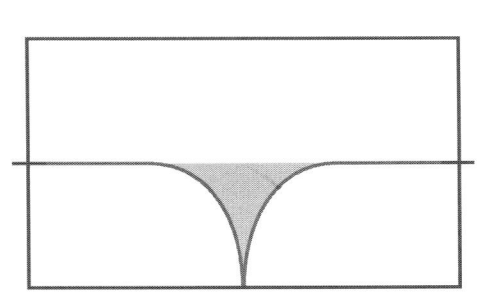

FIGURE 11.9
Notch-filter EQ curve.

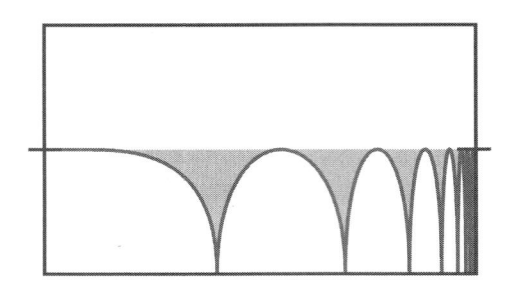

Audio 11.9 ▶). All frequencies except a particular area are let through by this filter.

11.2.5 Comb Filter

This filter is a series of notch filters that are harmonically related to one another (Figure 11.10; Audio 11.10 ▶).

FIGURE 11.10
Comb-filter EQ curve.

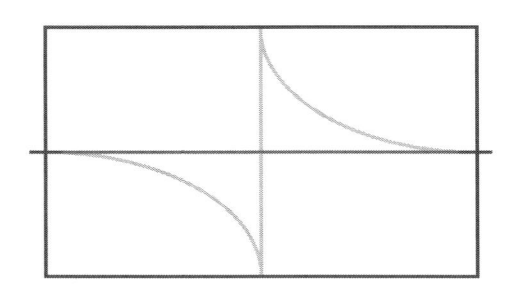

11.2.6 All-Pass Filter

This filter is used only to hear the filter effect (phase shifts) without removing any frequency content (Figure 11.11; Audio 11.11 ▶).

FIGURE 11.11
All-pass-filter phase-distortion curve.

The sound of a filter can also be accentuated by using the resonance control: a boost of frequencies around the cutoff point (Figure 11.12).

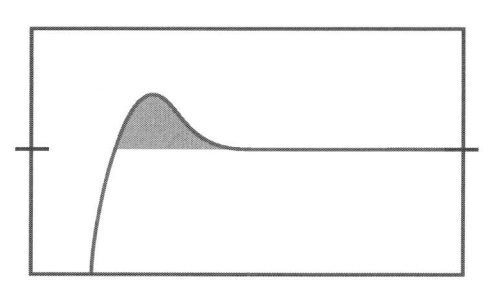

FIGURE 11.12
Resonant high-pass-filter EQ curve.

11.3 Modulators

In modular synthesizers (this also includes most subtractive synthesizers), modulators can be used to change the behavior of the different parameters available over time. The use of key tracking to modulate the oscillator's pitch is the most common example (which allows the synthesizer to generate a waveform at 261.6 Hz, or middle C, when the middle C key is pressed for example). Another common example is the use of an envelope to modulate the amplifier volume (which controls the volume of each note over time when a key is pressed). While different synthesizers will have different modulators and parameters that can be modulated, the most common ones include the following:

11.3.1 Envelope

An envelope is used to control particular parameters over time in a "one-shot" manner (i.e., it does not repeat). An envelope can be considered to be a way to automate a parameter by using the common attack-decay-sustain-release (ADSR) controls (Figure 11.13):

Attack: How quickly the parameter reaches the decay start position after the key is pressed.
Decay: How quickly the parameter reaches the sustain position once the attack time has passed.
Sustain: The position at which the parameter will remain as long as the key is pressed after both attack and decay times have passed.
Release: How quickly the parameter reaches the starting position after the key has been de-pressed.

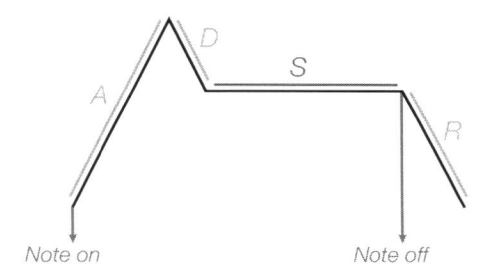

FIGURE 11.13

Attack-decay-sustain-release (ADSR) envelope.

11.3.2 Low-Frequency Oscillator

A low-frequency-oscillator (LFO) is an oscillator that vibrates at a frequency below 20 Hz, which is the lower limit of the human hearing range. Because of this limit, we actually hear the cycles of this oscillator instead of a particular pitch. LFOs are used as modulators to give repetitive or rhythmic effects. When an LFO is used as a modulator, its amplitude dictates the modulation amount, and its frequency dictates the modulation rate (Figure 11.14).

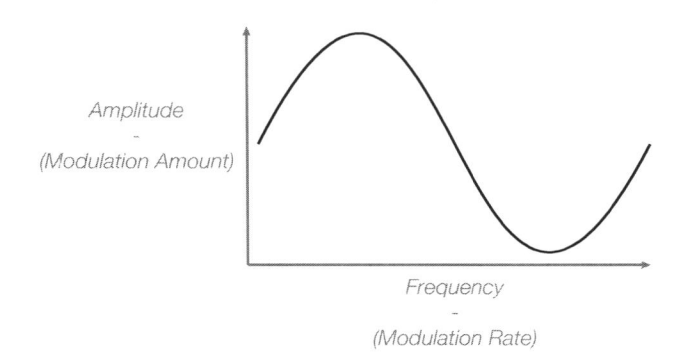

FIGURE 11.14

Low-frequency oscillator (LFO).

11.3.3 Key Tracking

This function will allow a parameter to move according to the key being pressed. Key tracking is almost always applied to pitch, since we want the notes generated by the synthesizer to follow the corresponding keys pressed on the keyboard.

11.3.4 Velocity

Since velocity is a MIDI function, its value (0–127) can be used to modulate any parameter of a synth. It is most commonly used to modulate the amplitude of the waveform: the harder the key is pressed, the louder the sound. Velocity can, however, be used to modulate pitch, filter cutoff, or any other available parameter. The same applies to any other available MIDI parameter, such as key tracking and after touch.

11.3.5 After Touch

After touch is commonly used to modulate the depth of vibrato on virtual string instruments. This effect emulates how real players move their fingers when bowing the same note for a long period of time.

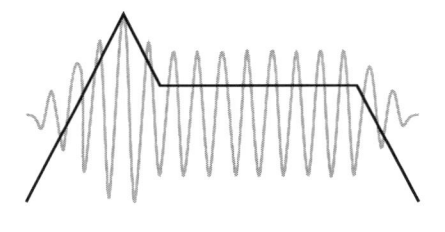

11.3.6 Modulation Examples

FIGURE 11.15
Envelope-modulating amplitude.

Using an envelope to modulate the amplifier (Figure 11.15; Audio 11.12), can use the ADSR values to control the volume of each note over time.

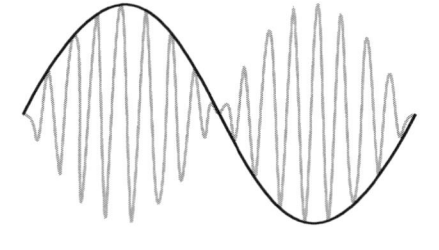

FIGURE 11.16
Sine-wave LFO modulating amplitude.

Using a sine-wave LFO to modulate the amplifier (Figure 11.16; Audio 11.13) is how tremolo effects are created.

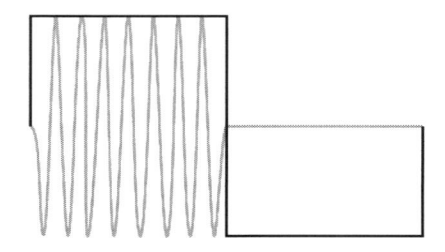

FIGURE 11.17
Square-wave LFO modulating amplitude.

Using a square-wave LFO to modulate the amplifier (Figure 11.17; Audio 11.14) is how stutter effects are created.

Using an envelope to modulate the pitch of the oscillator (Figure 11.18; Audio 11.15 ▶) makes the pitch start high and quickly drop to a lower value. This method is the basis for most drum synthesis.

FIGURE 11.18
Envelope modulating pitch.

Using a sine-wave LFO to modulate the pitch of the oscillator (Figure 11.19; Audio 11.16 ▶) is how vibrato effects are created.

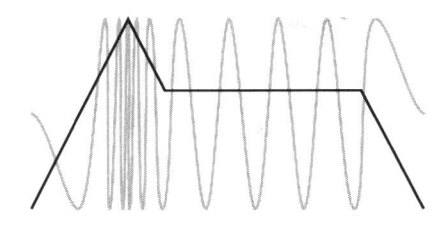

FIGURE 11.19
Sine-wave LFO modulating pitch.

Using an envelope to modulate the cutoff of a highly resonant filter (Figure 11.20; Audio 11.17 ▶) creates sci-fi laser sounds!

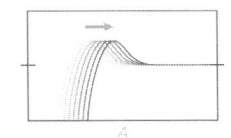

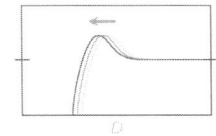

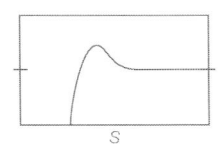

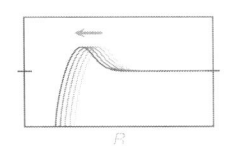

FIGURE 11.20
Envelope modulating high-pass-filter (HPF) cutoff.

Using a sine-wave LFO to modulate the cutoff of a bandpass filter (Figure 11.21; Audio 11.18 ▶) is how wah-wah effects are created.

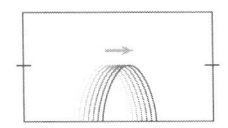

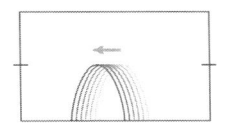

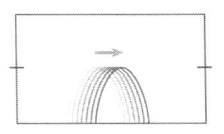

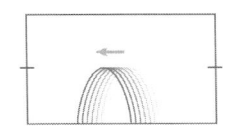

FIGURE 11.21
LFO modulating bandpass-filter (HPF) position.

Using a sine-wave LFO to modulate the cutoff of a LPF (Figure 11.22; Audio 11.19 ▶) is how wobble-bass sounds are created.

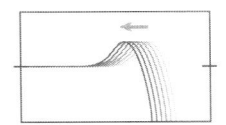

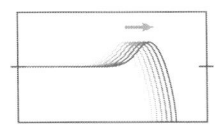

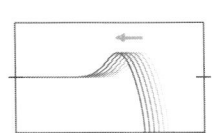

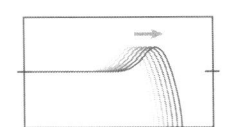

FIGURE 11.22
LFO modulating low-pass-filter (LPF) cutoff.

Using key tracking to modulate the cutoff of a filter (Figure 11.23; Audio 11.20 ▶) allows the cutoff frequency of the filter to follows the keys pressed. Since the "sound" of a filter is most prominent around its cutoff point, this function allows the filter sound to always be heard.

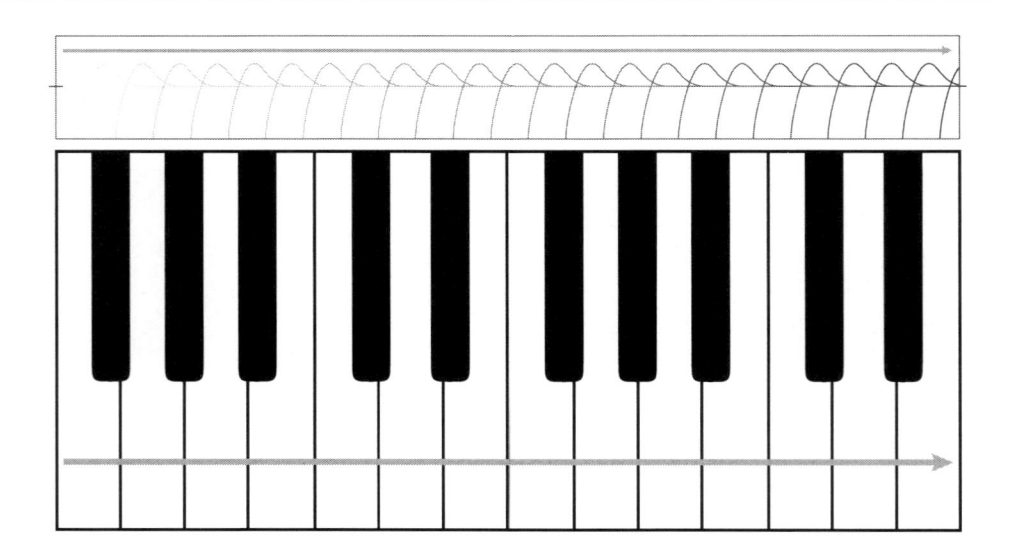

FIGURE 11.23
Key tracking modulating HPF cutoff.

11.4 Frequency-Modulation Synthesis

While subtractive synthesis is relatively easy to understand and the controls give predictable results, frequency-modulation (FM) synthesis is a lot harder to understand and predict. In short, FM synthesis works by modulating the pitch of one oscillator with the frequency of another one. In much the same way that modulating the pitch of the oscillator with an LFO gives us a vibrato effect, when the LFO oscillates faster than 20 Hz, it gives us frequency modulation. Effectively, FM synthesis is very fast vibrato (Audio 11.21 ▶). FM synthesis also often has a distinct 1980s tone (due to the popularity of the Yamaha DX7 at the time) and can easily recreate instruments with enharmonic partials, such as bells.

The modulator is often set at a fixed ratio from the key being pressed. This ratio follows the harmonic series to ensure that whole-number ratios work harmonically with the key being pressed. The volume of the modulator then acts as a "wet/dry" control for the amount of modulation (more modulation equates to more harmonics being added).

11.5 Amplitude and Ring Modulation

The same concept as FM can also be applied to the tremolo effect (Audio 11.22 ▶). Using a sine-wave LFO to modulate the amplifier gives us tremolo, but if the LFO oscillates faster than 20 Hz, the result is called amplitude, or ring, modulation. Depending on whether the modulation is unipolar (amplitude modulation) or bipolar (ring modulation), the original note played by the carrier will remain or disappear. The result often sounds like "metallic" distortion, something of a bit-crushing FX, or sample and hold FX. The oscillator used for ring modulation differs from FM synthesis in that its frequency is most often fixed rather than working as a ratio of the carrier frequency.

11.6 Instrument Design

Having a good understanding of the concepts explained in the last few subsections will allow you to use any subtractive synthesizer to create sounds. The key to virtual instrument design is to understand how the intended sound behaves in terms of frequency content and amplitude over time. As with every acoustic instrument mentioned in the previous chapters, choosing the right waveform is essential to reproducing a particular sound. It can also be useful to think of how a particular acoustic instrument generates sound. For example, when a violin is bowed, the strings are pulled in a linear fashion, and when the tension is too strong, they snap back to their original position before the process is repeated. Visualizing how the string moves uncovers that sawtooth waveforms are the closest shape to this movement. Another example is that of reed instruments, which basically offer two positions: stuck against the top or the bottom of their enclosure. This movement is similar to the shape of a square waveform. After that, choosing the right filters, amplitude envelope, and other FX will be needed to shape the sound.

With the loudness war raging and the need to have productions that always sound bigger, the current trend has shifted away from the complex waveforms and textures used in the 1980s and 1990s. Modern synth sounds often use simple waveforms to allow the final mix to be much louder and punchier. It is important to remember that the fewest number of instruments and sounds in the mix, the bigger each of them can be.

11.7 Enhancing Recorded Instruments

A basic knowledge of how synthesizers operate allows an engineer to enhance recorded instruments in subtle ways. For example, if a bass guitar lacks body and weight, converting it to MIDI by either transcribing the performance or using dedicated "audio to MIDI" processors will allow you to generate extra-low-end frequencies and layer them below the original bass recording. Being able to synthesize a kick drum and layer it with a recording can greatly expand the sonic palette available to an engineer to enhance the attack or body of the original sound. As another example, using samplers to layer extra strings below a recorded quartet can greatly thicken the performance while retaining a degree of realism from the original performance.

11.8 Programming Real Instruments

When you are working with sampled instruments such as string or brass sections, it is important to follow a similar arrangement to the accepted norms in composition. This guideline means that both the composition and the mixing need to feel realistic. For example, when you are programming an orchestral string section, each part needs to remain monophonic (as violins, violas, cellos, and double basses are monophonic instruments), but the polyphony can be created through the different instruments combining together. In this case, the first violins will harmonize with the second violins, the violas, cellos, and double basses for example.

The mixing of these instruments will also need to recreate what would happen on a real stage. Thus you would pan the first violins left (using both intensity and phase stereo to pan more authentically) and you would make them feel close to the listener by using high volume, low reverb amount (more on this subject in Chapter 16, "Reverberation and Delays"), and careful use of midrange EQ to bring them forward in the mix. You would mix the second violins very similarly, but pan them middle left. The violas would also be similar but panned middle right. The cellos would be panned right and a little closer to the listener, though with less thickness, as there are fewer of them, and the double basses panned right and placed behind the cellos (Audio 11.23 ▶).

Chapter 12

Mixing

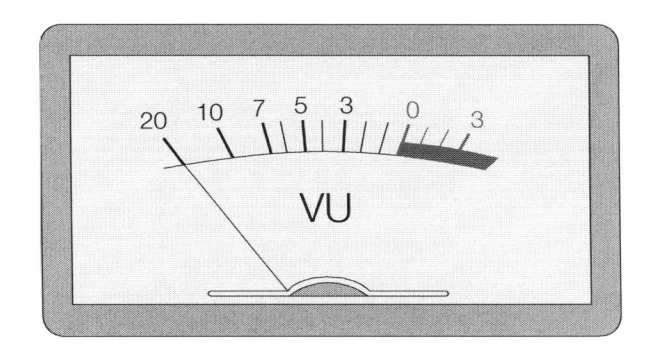

Mixing takes place after all parts of a song have been recorded and edited. This task can be extremely easy when you are working with quality recordings, or extremely frustrating when you cannot achieve the desired outcome. One important point must be made before discussion of the technical aspects of this task: mixing is all about listening. It is about making the hook strong, and embellishing the other elements to the extent where they support the hook. The aim is not to use every tool and technique on every mix, but instead recognize what the mix needs and use the tools at hand to "fix the bad stuff and enhance the good stuff." In the digital world, it is very easy to over-process tracks with dozens of plug-ins that, if you believe the hype, are all "revolutionary." The truth is that these are only tools, and knowing when to leave them out is just as important as knowing when use them.

The best tools used for mixing are your ears and monitoring environment. No amount of expensive equipment will ever be as useful as experience and good-quality monitors in a well-treated room. There is no "magic box" that big-name engineers use, only their experience and ability to hear the music properly.

12.1 The Mix as a Virtual Soundstage

Every recorded song should give the illusion that it is being played on a virtual stage. As such, there are three dimensions available when placing instruments within the stage (Figure 12.1):

- X: This dimension is the easiest to understand and relates directly to the stereo spread between the speakers.
- Y: While this dimension may not be obvious at first, it relates to the frequency content of each instrument, as bass is made up of low frequencies and treble contains high frequencies.
- Z: This dimension is the front-to-back relationship of the instruments within a mix, and the illusion of depth of field (or depth of the virtual sound stage).

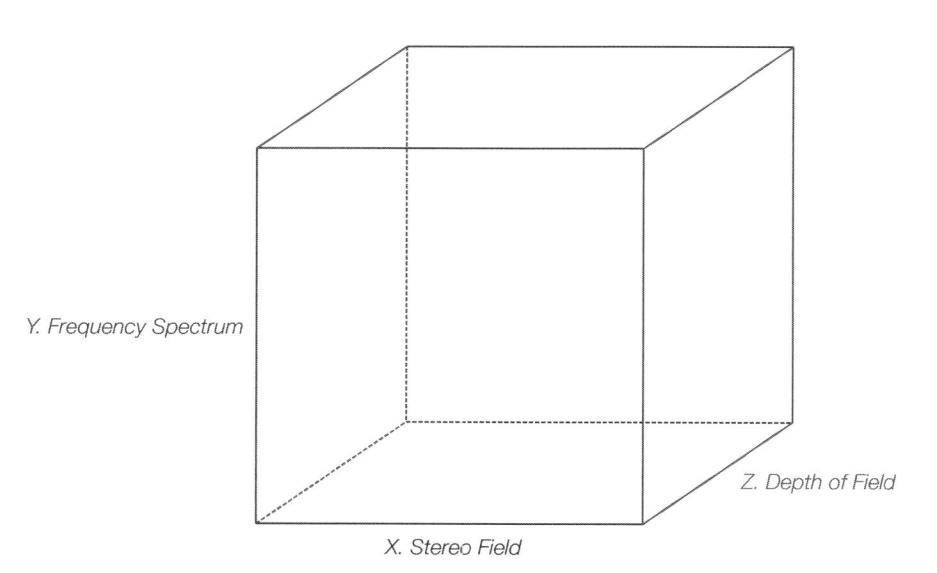

FIGURE 12.1
Virtual soundstage dimensions.

The main rule to follow when placing instruments within our three-dimensional soundstage is that no two instruments can occupy the same position at the same point in time. The same principle applies to real-world objects; two objects cannot be placed at the exact same location at the same time! It follows that a mixing engineer must use the tools at hand (panning, EQ, volume, reverb, etc.) to ensure that no two instruments occupy the same frequency areas, are as loud as each other, and are panned the same way.

Once all of the instruments have been fit into the soundstage, the mix is something of a "picture" of the song. Since a picture is static, the next step is to create interest and engage the listener with the song and the mix, and turn this picture into a movie. You can do so in a number of ways at various stages of the production, but when you are dealing specifically with the mix, at this point automation and micro-modulations come into play. Automating parameters such as volume, EQ and panning will allow for the instruments to move within our 3D soundstage, thus turning our "picture" into a "movie." This step is sometimes considered to be an art in itself, since the elements of the mix must move with the song as if they were all skillfully dancing with one another.

It is also important to note that the shape and size of the virtual soundstage (or the "shape" and size of the cube above) is determined by the reverb used in the mix—more specifically, the early reflection part of the reverb (for more on this topic, see Chapter 16, "Reverberation and Delays").

12.2 Mixing Workflows

Mixing is often referred to as being both an art and a science. To practice both, audio engineers tend to learn the rules first (the science), only to be able to break them to leave room for their own creativity (the art). While all engineers have their own methods and views on how a mix should be approached, the following four workflows can provide good frameworks by which to work through a mix. Note that these approaches are only "mindsets" that can help set a clearer path toward a great-sounding mix. How different engineers use, adapt, and ultimately create their own ways of working is what sets everyone apart.

The first workflow sees mixing as a fluid process that starts in the left brain (the technical side), and slowly moves toward the right brain (the creative side). This approach can prove to be very effective for engineers who are rather methodical in their ways of working. It follows a clear path that runs from fully technical aspects of mixing such as strip silence and gain staging to fully creative ones such as enhancement reverb and automation. Since a lot of techniques given for this workflow will also apply in the following ones, this section of the chapter will be larger than the remaining three.

The second workflow tackles mixing from a slightly different viewpoint. It assumes that the mix quality is limited by the engineer's skills, and tries to approach mixing in a similar way to how those skills are acquired. This "bottom-up" approach can be effective at fully using the current set of skills that the engineer has, while reducing the issues associated with the missing skills. By clearly identifying what is lacking in an engineer's skill set and slowly working toward filling those gaps, this approach can help students to accurately self-reflect and therefore speed up their learning process. Note that the order of processing elements in a mix using this workflow is very similar to the previously mentioned left brain/right brain approach.

The third workflow, and arguably the hardest to implement, works in the opposite way: "top down." When the critical listening skills of an engineer are developed to a point where he or she can see "the big picture" right from the start, this approach becomes the most effective at using a minimal amount of processing and achieving the best possible quality. Overall, being able to work this way should be considered the "holy grail" for most engineers, although it can only be assimilated organically through years of practice in critical listening and mixing.

The fourth workflow, "foreground versus background," is an approach that categorizes the elements of the mix in order of importance, and suggests the appropriate techniques for each. While it can often be seen as a simpler way to mix a song, you should be careful to keep this method only as a backup plan when others have failed to provide good results, rather than using this workflow systematically. Because of its simplicity, it can easily fool the engineer into settling for mediocre results if every mix is approached this way.

Overall, and no matter which workflow is being used for mixing, the tools and concepts overlap in order to attain a common goal: a great-sounding mix. They simply provide different perspectives that may be more appropriate for different kinds of engineers.

12.2.1 Left Brain/Right Brain Approach

This mixing approach starts with the technical side of mixing and slowly moves toward the creative side. This technique can also be seen as moving from individual instrument processing to the whole mix in one fluid motion to ensure that the creative flow builds and is not disturbed throughout the mixing process. The first part involves cleaning audio tracks and does not require any particular skills, only attention to detail in what could ultimately help with headroom and separation. From there, the balance, EQ, panning, and reverb parts have both technical and creative aspects, jumping back and forth between the different sides of the brain. The last step is creating movement within the mix, a highly creative task that aims to see the engineer as an extra band member, with the mix as his or her instrument.

12.2.1.1 *Cleaning Audio Tracks*

Strip Silence

The first step is to delete audio clips that do not contain any sound (Figure 12.2). Quite often, engineers keep recording between different parts of songs even if the musicians are not playing. The result means that there are areas where "silence" is recorded. This "silence" is actually very low-level noise, not digital silence. Deleting all those parts will ensure that the noise floor is kept as low as possible and will ultimately give the mix a wider dynamic range.

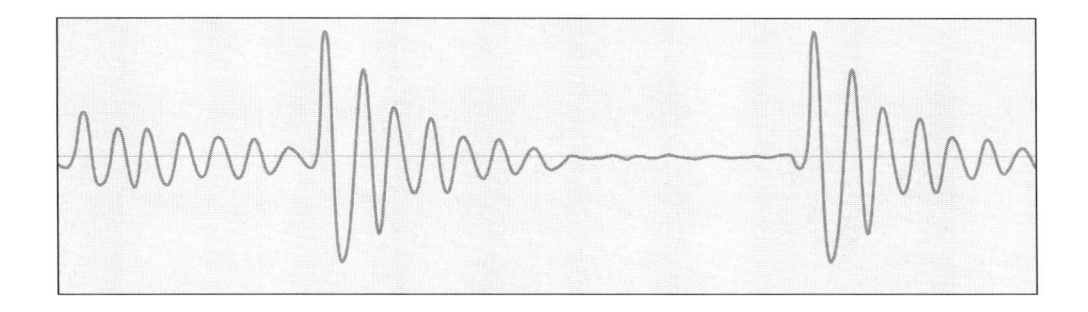

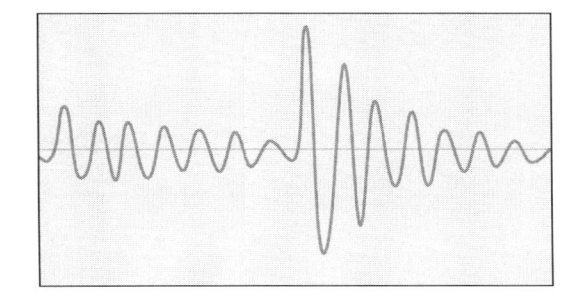

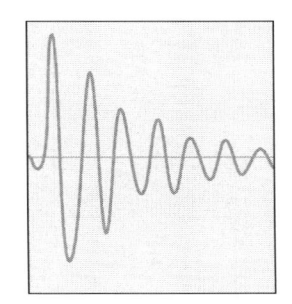

FIGURE 12.2

Stripping silence.

Gain Staging

The next step is to set the volume levels for all the tracks in the session. This step is important for ensuring that no plug-in distorts on input. Assuming that the session is at 24 bit, all tracks should peak around −18 dBfs before any plug-in is inserted. Be careful not to use the volume fader of the track for this task, as this control is often post-plugins. Working this way allows for more digital headroom and only "wastes" 3 bits of audio (1 bit = 6 dB and 3 × 6 dB = 18 dB). When you are working at 24 bit, this rate is still above the average analog equipment and the human ear's dynamic range (roughly 120 dB or 20 bit).

Filters (High-Pass Filters and Low-Pass Filters)

Following the same principle as for strip silence, some engineers use HPFs on most tracks except kick and bass. Frequency analyzers can often accurately display where the fundamental frequency is at any given point. Ensuring that the filter is set below the lowest fundamental played by the instrument is all that is needed. More often than not, using a steep slope such as 48 dB per octave works fine, but there are times when such a slope can cause phasing issues in the leftover frequencies. This problem is especially true for bass-heavy instruments. If you have such issues, using a smoother curve is necessary. Even though frequency analyzers can tell approximately how high to set the cutoff point, listening to the track in solo ensures that the sound remains "intact" and that only inaudible low-end rumble is removed.

Note that while this technique works well for large mixes with a lot of tracks, it becomes less necessary as track count lowers. High-pass-filtering everything may make the mix sound sterile and lifeless. It is important to always listen to the overall "vibe" of the mix after this stage to ensure that the life has not been taken away. If this is the case, backing off the filters and using more gentle slopes can bring life back into the mix.

Depending on the type of track being worked on, adding LPFs to clean up the high end of the spectrum can also be useful. This step is not as critical as using HPFs, but can sometimes ensure that the recording is not too harsh (Figure 12.3).

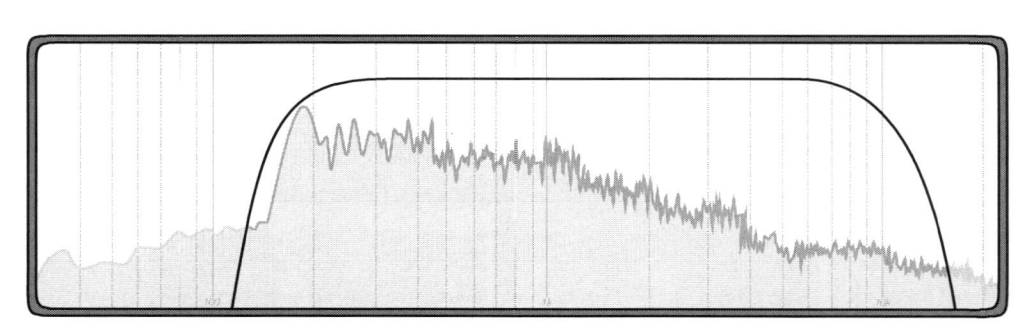

FIGURE 12.3
High- and low-pass filters to clean up unused frequencies.

Correlation

The next step is critical to ensure that no stereo tracks have fully out-of-phase signals that will cancel each other when switched to mono. Correlation meters (Figure 12.4) are used to show how a stereo track will behave when summed to mono. Keeping the meter roughly halfway on the positive side is often a good compromise between phase cancellation and stereo width. Generally speaking, if the correlation meter is in the negative side, the volume of the instrument will lower in mono, and its tone will change. If the correlation meter is in the positive side, only a change in tone will result.

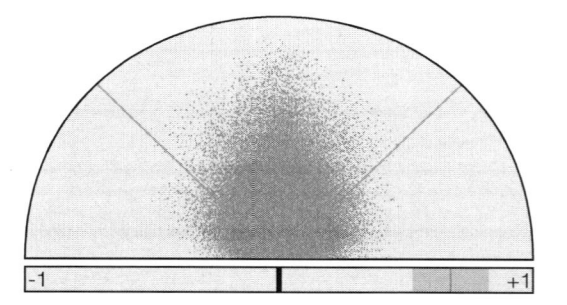

FIGURE 12.4
Correlation meter.

There are various different techniques that can be used to bring a signal more in phase, but only trial and error can reveal which is the most effective. On multi-microphone recordings such as that of a drum kit, it is preferable to time-align the different elements manually (or by using phase-alignment plug-ins), as this alignment will not affect the stereo width, but only the frequency content. In some cases, where the phase displacement is randomized (such as in organs recorded by using a Leslie cab), using a mid-side EQ to remove low frequencies by using a steep HPF from the side channel can work well. This method should be preferred if the cutoff frequency can be set lower than 500 Hz, as the track will retain most of its stereo image. This technique works, as it is often the low frequencies that get out of phase because of the length of their waveform. The last tool that can be used is the pan control. Reducing the stereo width of a track gives a truer representation of how the track will sound when switched to mono, although this reduction will ultimately affect stereo image the most.

As previously mentioned, even when the correlation meter is in the positive side, the tone of an instrument may be affected when you switch to mono. This is one case when using mid-side processors such as EQ and harmonic enhancers, or "re-injecting" the side signal into the center channel (and vice versa), can be useful to "match" the mid and side tone (for more on this topic, see Chapter 18, "Mastering").

12.2.1.2 *Rough Volume Balance*

Although the volume of each track is often modified at every stage of the mix, ensuring that every instrument can be clearly heard at this point is critical. This step sometimes involves doing volume or trim-gain automation for very dynamic parts.

12.2.1.3 *Panning*

The next step of the mixing process is to achieve as much separation between the instruments as possible by using only the X-axis of our virtual soundstage. Limit yourself to using panning only to achieve as much separation as possible. As this can be a hard step to undertake, the tip laid out in Box 12.1 can help facilitate this process. This step will ensure that this dimension is used better than when you are relying on other processes to help separate instruments. It is important to keep in mind that the mix should be "frequency balanced" across the stereo field to ensure that a high-frequency sound panned to one side is counterbalanced with another high-frequency sound on the other side, for example.

> **BOX 12.1**
>
> *Spend time separating instruments with panning, as it will ultimately result in a larger-sounding mix! If you are having trouble making everything fit together at this point, try muting all the elements that will be panned in the center, such as lead vocal, kick, snare, and bass. This muting may help free up the center image of the mix while other elements are positioned. Once you have achieved a good separation between the remaining instruments, un-muting the center instruments will not change how the rest should be panned, but how far to the sides they should be panned.*

12.2.1.4 Equalizing and Compression

The next step in the mixing process is to switch the monitoring to mono. In much the same way that a vision-impaired person often has a more developed sense of hearing, reducing the available dimensions for placing instruments in the virtual soundstage will ensure that the remaining ones are better used. Force yourself to keep the mono switch on until as you have achieved as much separation as possible by using EQs. Following the same principle, do not jump to the next step and use reverbs until all the elements of the mix have been equalized.

Before separating instruments on the Y-axis by using EQs, a little preplanning is necessary. It involves mapping where the instruments will fit in the frequency spectrum. Where the instruments are placed will be largely dependent on the genre of music and the number of instruments in the mix. In order to fit instruments around one another, use both cuts and boosts. For example, when trying to fit kick and bass around one another, you could boost 60 Hz and cut 200 Hz on the kick, while cutting below 60 Hz and boosting 200 Hz on the bass (Figure 12.5).

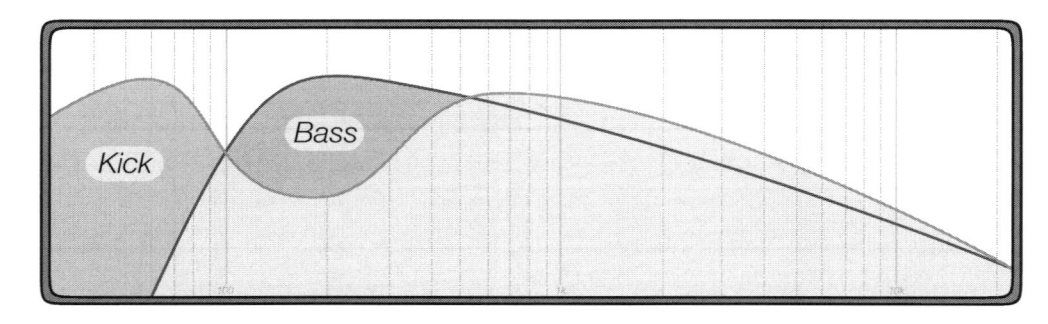

FIGURE 12.5 Kick and bass placement within frequency spectrum.

This process is referred to as "complementary" or "niche" EQ, in which each instrument has a niche frequency that it "lives" in. The main goal is to achieve maximum separation and ensure that the instruments do not "mask" one another. Note that above 10 kHz, separating instruments becomes less critical, as high frequencies mostly contain partials (harmonics and overtones) and no fundamentals.

While it is commonly accepted that equalizing should be done when the whole mix is playing rather than when you are soloing each instrument, it is often hard to work this way. One way to get around this difficulty is to "work up" to equalizing with the whole mix playing. The following technique can help get to that point by scaffolding EQ moves. It works in three steps:

Fixing and Separation Equalizer plus Compression (Each Instrument Soloed)

This first step involves working with EQs to remove unwanted frequencies that should not have been recorded in the first place, including resonant frequencies in hi-hats or the "boxy" characteristics of a kick drum. The preplanned EQ curves to make instruments fit around one another should also be implemented at this stage. At this time you should compress instruments individually, since compression can change the tonal aspects of an instrument and therefore affect how it can be fitted around others.

Separation Equalizer (Instruments with Similar Frequency Content Soloed)

Once all the instruments have been equalized in isolation, it is time to adjust these EQ curves on instruments with similar frequency content. An example would be to listen to the kick and bass together while you are adjusting the EQs on both, then once maximum separation has been achieved, mute the kick and listen to the bass and guitars, repeating the process. It is important to work from bass to treble, as this order is how masking occurs naturally; sub-bass masks bass, bass masks low mids, low mids masks mids, and so on.

Final Equalizer Adjustments, Including Enhancement Equalizer (All Instruments Playing)

The final step is to play the full mix and do the final EQ adjustments (Figure 12.6), which should be minor and should not greatly affect the tone of any instrument within the mix. These EQs may need to be slightly readjusted after reverb and other audio processing are added.

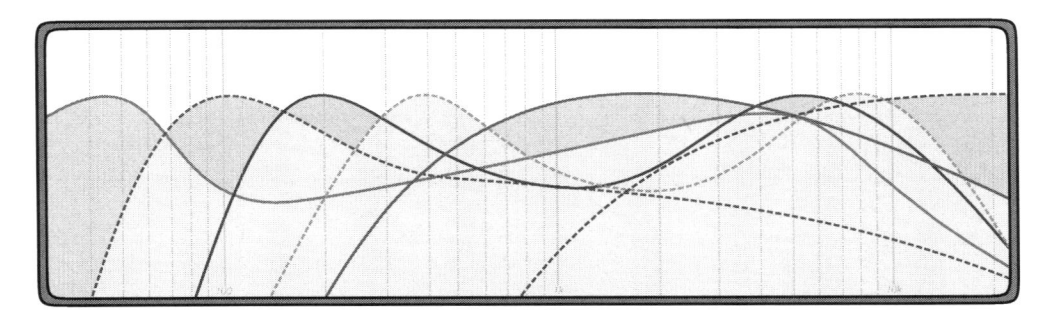

FIGURE 12.6

EQ curves to fit various instruments within frequency spectrum.

It is a good idea to keep a correlation meter open while equalizing, as the different EQ types (linear phase, minimum phase, etc.) will have an effect on the phase relationship of the different elements of the song. This effect can influence mono compatibility when you are equalizing stereo tracks.

It is also important to try phase-alignment tools rather than EQs for low-end instruments, as these tools can often achieve great results without the use of extra processing. Do not forget that these tools work only when all instruments in the session are playing, and that the volume of each instrument will also affect how different instruments affect one another in relation to their phase.

Once you have done this part and achieved the most separation by using EQ and compression in mono, it is important to return to stereo monitoring and check that the overall tone of the mix is similar. Careful use of mid-side equalizing can help maintain consistency between the mono and stereo mix. For example, if a sound is duller in mono than in stereo, brightening the center channel while making the sides duller on that track will work. The overall tone should remain the same in stereo, but the mono version will now be closer to it. The opposite can also work if the sound is brighter in mono (although this occurrence is rare). Using mid-side EQs on single instruments should not affect the separation previously achieved unless the separation had been done by using mid-side equalizing to begin with.

From here, the rest of the mix should be done while you are monitoring in stereo, although you need to do regular checks in mono, as reverberation and micro-modulations can affect how the mix translates in mono.

12.2.1.5 *Reverb*

Once all the instruments have been equalized and compressed, the mix should feel balanced and controlled. Now is the time to add another dimension to the mix: depth (via reverb). This step should always be left until the very end of the mix. Everyone loves reverb; it makes everything sound nicer and more polished, but it also creates mud and blurs the lines between instruments. Since reverbs will destroy some of the separation achieved by using EQs, it is important to use them sparingly and only after equalizing. Adding reverb after EQ means that you already know how much separation can be achieved and will be able to hear the point at which the reverb starts to harm the mix. Working this way will reveal how little reverb is actually needed in the mix and how much extra punch can be achieved. When judging reverb level, ensure that the dry signal is not masked when checked in mono.

The texture and quality of the reverb should be adjusted while you are listening in stereo, but the amount of reverb applied should be set while you are monitoring in both mono and stereo. Both placement and enhancement reverbs should be done at this point (for more on this subject see Chapter 16).

Never forget that when it comes to reverb, less is more. You are adding the icing on the cake, not turning the whole song into an icing cake!

12.2.1.6 *Movement*

Now that you have achieved a solid static mix, it is time to make it move! The aim is to craft a listening experience that is interesting at all time and reveals new elements with each play. In much the same way that looking at a still picture for three minutes would get boring, a mix without movement can feel very stale by the end of the song. On the other hand, if every element of the mix has movement, the interest is created by making the listener notice new elements with each new play of the song. Movies work in a similar way. Watching a movie for the first time, you will understand the main plot and notice a few small elements of particular interest. When watching it a second time, you may notice more elements, such as a pun that could have gone unnoticed and notable streets signs in the background. Some of the movement in the mix will already be present through dynamic performance of the instruments during the recording as well as the use of the different tricks mentioned in the recording chapters.

The main tools used for giving life to the mix are automation and micromodulations. Depending on the size and complexity of the mix, a full-day session in the studio may be needed to achieve the necessary movement. Tempo changes can also give extra movement to a song (adding one or two beats per minute (BPM) to the chorus for example), but should be done during the recording stage, as it is nearly impossible to change the tempo without audible artifacts after everything has been edited and mixed.

Automation

Automation is the main tool for making the different elements of the mix move within the 3D soundstage. As such, automation should include the following:

- Panning, EQ, and Haas effect delays for the X-axis
- EQ for the Y-axis
- Volume, EQ, and reverbs for the Z-axis

Being very efficient with your DAW means that this step can be done quickly and does not feel daunting. Many audio engineering students skip this step, as it takes them hours to implement because of the lack of efficiency within their DAW of choice.

Automation should be done section by section (verse, chorus, etc.), but it should never be copied from one verse or chorus to the other. Treating each part with a new ear means that the song develops differently over time. At a minimum, it is essential to use automation to change the frequency content of different instruments depending on how many other instruments are playing simultaneously. For example, if the verse contains only guitars and vocals, but the chorus has bass and drums added, the EQ curves of the guitars will need to change to accommodate the extra instruments during the chorus. Similarly, the amount of reverb present will need to be adjusted depending on how much space is available in the mix. Sparse sections will allow for more reverb, while denser sections will require less reverb. Here are a few examples of common automation moves that can also help the mix move forward:

- Reduce the overall stereo width toward the end of a verse and return to full width at the start of the chorus. The chorus will automatically feel wider without your having to use extra processors.
- Reduce the frequency content of sustained instruments such as guitars or synth pads toward the end of a verse, and then return to full bandwidth at the start of the chorus. The chorus will feel fuller and have more impact.
- Increase the volume of the first note or hit of the chorus. This step will give the impression of a louder chorus without actually changing the overall volume.
- Increase the overall reverb toward the end of a verse, and then return to a dryer setting at the start of the chorus. This move will give the chorus more impact and punch.
- Use different combinations of the four automation moves previously mentioned (width, bandwidth, volume, reverb) (Audio 12.1 ▶).
- Program random automation on volume, EQ gain, and frequency of a bell boost on each hi-hat or snare hit to create subtle movement in the performance (Audio 12.2 ▶).
- Increase the brightness of some instruments during the final chorus of the song, giving the impression that the mix is getting fuller.
- Increasing the level of drum fills or guitar licks to ensure they are heard louder than other instruments.

Automation is also used to swap "stars" of the mix at any given point. Since the listener's attention should mostly be directed to one element at a time, volume automation can be used to bring an instrument to the listener's attention and then send it again to the back of the soundstage.

There is also a phenomenon of lasting mental image, which happens when an instrument is brought to the front of the mix for a short period of time (by raising the volume of a drum fill or guitar lick, for example). Once the instrument has been pushed back in the mix, the listener's attention is still focused on it. This experience is similar to looking at a nice car pass by and get farther away. Even though it cannot be properly seen anymore, the person watching it is still focused on it.

Micro-Modulations

Micro-modulations are a type of automation that can give a lot of movement to the mix if used correctly. Be careful not to overuse these tools, as they can be very distracting if the listener notices them. Using modulation plug-ins such as flangers, auto-filters and other evolving audio FX, you can create movement on particular tracks without having to rely on automation. It is preferable to set these processors to sync with a musical value of the song's tempo. Here are a few examples of common micro-modulations:

- Flanger or phaser on hi-hats (if using a phaser, ensure that it resonates at a note that works with the key of the song) (Audio 12.3 ⏵)
- Auto-filter on snare (Audio 12.4 ⏵)
- Auto-filter on rhythm guitars (automating the resonance of the filter can add even more movement) (Audio 12.5 ⏵)
- Auto-pan on pads (Audio 12.6 ⏵)
- Chorus on reverb returns (this micro-modulation is most useful on reverb tails used for non-fretted instruments) (Audio 12.7 ⏵)

12.2.2 Bottom-Up Approach

Becoming a great mixing engineer takes time and practice. Over the years, different skill levels will be achieved that follow a similar sequence for most engineers: leveling, separation, coherence, and interest. The "bottom-up" approach to mixing follows this sequence to ensure that every aspect of one skill level has been implemented into the mix before moving on to the next. The first three steps require the balancing of different aspects of the instruments, while the fourth step is the sweetening of the mix.

12.2.2.1 *Leveling*

This step is the simplest, although most important, aspect of mixing. The two tools available for "coarse" leveling of instruments are gain staging and volume faders. As mentioned earlier, gain staging should be done first, as it relates to using the DAW in a similar way to analog consoles, and then volume faders can be used to ensure that every instrument is heard properly. Spending a lot of time on this first step is crucial in attaining the best-sounding mix possible. Since not all instruments may be playing in every section of the song, some volume automation may be necessary at this point to rebalance the mix, depending on the section of the song. Note that panning forms part of this leveling process, as it is leveling of individual speakers. Panning can then be further refined in the next step. Overall, the keyword for this step is "balance"!

12.2.2.2 *Separation*

This step may be better understood as "micro-leveling" of instruments. Using EQs, which are essentially frequency-dependent volume controls, you need to level the different frequencies available in the mix. By ensuring that no two instrument overlap each other in terms of frequency, you can achieve a "finer" leveling of instruments. In this step, it is important to understand that while EQs are applied on individual instruments, the engineer should really be looking at "the big picture." This attitude means getting into the mindset of using EQs to further refine the leveling previously achieved and seeing these tools as a way to give an even frequency response to the mix overall. Oftentimes, an instrument that cannot be balanced by the use of its volume fader only (it is either too loud or too soft, but never quite right) is a good indication that it needs to be balanced against others with the use of an EQ. The main problem to listen for in this step is frequency masking. Do instruments feel as if they were sitting on top of one another, or do they have their own space in the mix? Once again, automation should form part of this process, as not all instruments are present in every section of the song. By changing the frequency content of instruments during the different sections, you can keep the mix balanced frequency-wise. Panning also overlaps here, as different frequencies need to counterbalance one another on either side of the stereo field. Overall, the keywords for this step are "frequency balance"!

12.2.2.3 Coherence

This next step is a continuation of the first two and has three main goals overall: individual instrument coherence, overall instrument coherence, and virtual soundstage coherence. The tools used in this step are mostly compression and reverberation (the early reflections part; for more on this topic, see Chapter 16).

At an instrument level, it is important that the instrument "makes sense" to the listener. For example, if the hi-hats of a drum kit are very dry and piercing, but the snare is more present in the overhead microphone (pushing it slightly further back in the mix), the kit itself will not have coherence. In this example, this step should have been taken care of at the leveling stage but is often overlooked, as the kit is often seen as being one instrument rather than multiple drums. Using a combination of reverb to push the hi-hats back, and overall compression on the kit, you can bring each element "in line" with one another and give overall coherence at the instrument level. Another example could be lead and backing vocals. Once again, these elements need to be balanced and processed to form one whole. Once again, slight reverberation and compression can tighten them up.

At the mix level, all the instruments need to "make sense." Similar to the drum kit example, if the guitars have a lot of reverb and are pushed back in the mix, but the drums are dry and at the front of the mix, the listener will be confused into thinking that the band is not performing on a realistic virtual soundstage. Once again, careful use of reverberation and bus compression can bring all of the instruments in the mix in line and give coherence at the mix level.

An extension of this process involves ensuring that the virtual soundstage created makes sense. By using the same ER reverb on all instruments, you can give the listener a clear picture of where the band is playing. The type of reverb to be used should have been decided during the preproduction process, as this aspect is something that can be refined during the recording process. Overall, is the band playing in a small club, a large stage, a cave? It is important to take time to critically listen to the overall mix and add reverberation as appropriate to glue the different elements together.

Once again, the keyword for this step is "balance"!

12.2.2.4 Interest

This last step is when an engineer can really let his or her creativity run free. At this point, all of the elements are balanced and fit together well; it is now time to sweeten the mix and give it a life of its own. The main tools used here are reverb (the tail part; for more on this topic, see Chapter 16), automation, micro-modulations, and other creative effects. Give two seasoned engineers a well-balanced mix, and this is the step at which they will steer it in different directions. The only rule to follow here is not to alter the balance previously achieved but instead support and sweeten the mix, hence the keywords for this step: "first, do no harm"!

12.2.3 Top-Down Approach

The top-down approach is the holy grail of mixing skills. You can think of approaching (and successfully completing) a mix this way as the ultimate Zen technique. This method is assimilated organically by veteran engineers through years of practice in critical listening, recording, and mixing. Ultimately, this approach is all about balance. Engineers need to have an ear that is developed enough to be able to listen for issues, and instinctively use the best tool in their arsenal to fix those problems. Overall, only the bare minimum of processing is used in each mix. This approach of mixing will grow organically with years of experience as you develop your critical listening skills and learn the tools of the trade.

12.2.4 Foreground versus Background Approach

A slightly different approach to mixing involves thinking of the mix as having two main parts: the foreground elements and the background elements (Figure 12.7). Using this approach can sometimes fix an otherwise "un-mixable" song. The foreground elements should be the most important and obvious parts, such as vocals, kick, snare, and guitar. These elements should come across as being clean and powerful, with a lot of separation and space between them, to help keep the overall power of the song going by retaining their weight and punch. Quite often, these elements contain a wide frequency range, are compressed to remain at the front of the mix, are dry in terms of reverb and FX, evolve quite rapidly through automation, and generally are the elements that listeners focus on when listening to a song. The background elements, on the other hand, should have less power, more FX, and slower movements to them, and fill the song in a way that may increase background blur by creating a "backdrop" to the entire song. While this way of approaching a mix may not always work, it can save an otherwise dull and uninteresting mix by putting forward the most important elements on top of a blurry musical background.

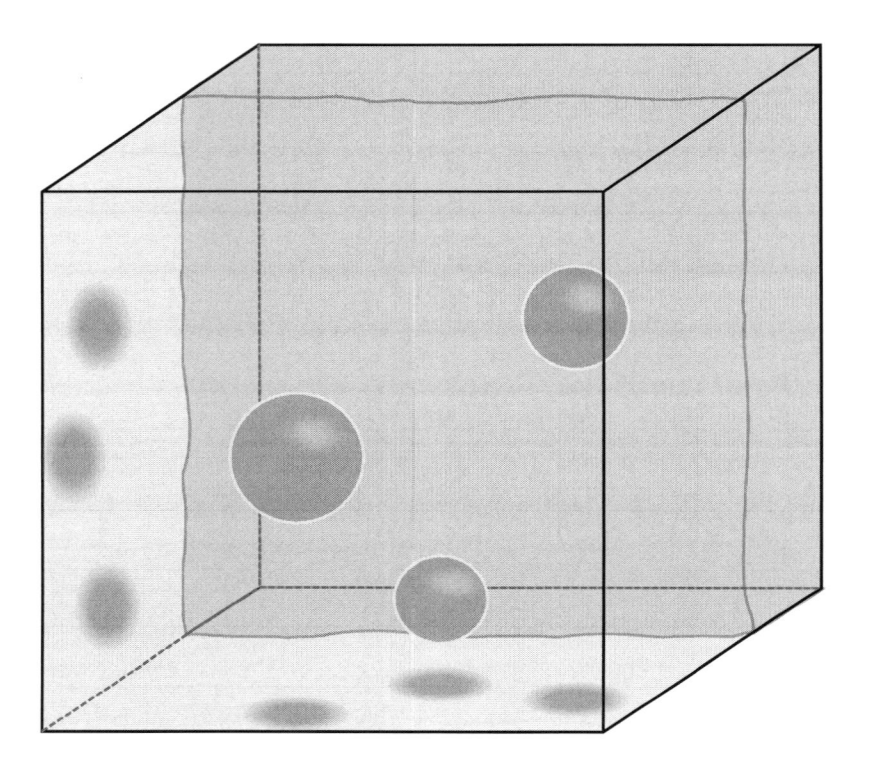

FIGURE 12.7
Foreground versus background virtual soundstage.

12.3 Monitoring Level

Just as mixes should be checked on different monitoring systems, they should likewise be checked at a variety of volume levels. Lower volumes are great for setting levels of transient heavy instruments such as drums, while higher volumes can help accurately set bass levels. This reaction is because our ears tend to ignore sharp transients when they are too loud. It is a "self-protection" mechanism that keeps ears from aching whenever they hear a brief loud sound. In much the same way, we have a distorted perception of bass when the volume is too low.

In a rule of thumb, monitor at the lowest "comfortable" level for most of the mix. This level should not be above around 80 dB (the monitoring level generally recommended for mixing). When you are judging dynamics and frequency content critically, checking at very low and high levels will help you make more objective decisions. Another advantage of monitoring at lower levels is that the room interacts less with the speakers, an effect that can make a large difference when you are working in non-acoustically treated environments.

12.4 Reference Tracks

The use of reference tracks is very effective in obtaining the intended sound. Knowing that an artist wants to sound like "band X" will make the mixing engineer's job a lot easier. Reference tracks are used at all stages of the production, from composition to mastering. For example, you will often hear a song that is very similar in arrangement, structure, and sound quality as another that is currently popular. This similarity is usually because the writer used a reference track while composing. Using references in mixing is also necessary to obtain a quality product and can greatly help in uncovering, for example, how much punch or brightness a mix should have or what the reverb levels should be. Using reference tracks can be a great way of "equalizing your ears." After only a few minutes of mixing, your ears can get accustomed to the sounds you are hearing, making them the norm. It is important to "reset" the ears once in a while to ensure that you do not lose perspective of the end goal. While taking regular breaks from mixing is crucial in being able to mix objectively, using a reference track is just as important in that regards.

12.5 Different Perspective

It is often said that a mix should never be completed on the same day it was started. Coming back with fresh ears on another day gives a fresh perspective and allows the engineer to hear aspects that may have been overlooked due to ear fatigue, become too familiar with the quirks of the mix, or simply have gone unnoticed. If you are pressed for time when delivering a mix and need to reset your perspective, a good trick is to call someone in the studio and play him or her the mix. That person does not need to be an engineer or a musician; he or she just needs to be there to listen to the mix. The simple fact that the engineer is playing back the mix to a third party often makes him or her quite critical over aspects that may have previously gone unnoticed. While this step may seem a little unusual, it works every time in getting in a different mindset when the engineer is listening to the same mix!

12.6 Target Audience and Intended Playback System

The target audience and intended playback system have influenced music composition throughout the ages. As David Byrne mentioned in his book *How*

Music Works (San Francisco: McSweeney's, 2012), particular genres of music make use of composition and arrangement techniques that suit their listeners and the environment in which the music is consumed. For example, music intended to be performed in large cathedrals does not contain many percussive elements that would get lost in reverb in that environment. Similarly, genres like early hip-hop have developed out of the need to create looped sections of songs so that dancers could enjoy themselves in clubs.

With recorded music and the manner by which it is mixed, it is still important to keep in mind the sort of playback system and the audience. Mixing music for an older demographic will be different from mixing for children because of the general quality of each person's hearing. Mixing a song for a TV commercial is vastly different from mixing for nightclubs because of the size of the speakers being used. The amount of compression used on classical music (made to be listened on high-quality playback systems in quiet listening environments) will be different from that used on top–forty commercial music (made to be listened on earphones; in cars, clubs, shopping centers; and in and other listening environments where the music must fight with outside noises). For classical music, using very little compression is acceptable, as the low-level sounds introduced by the high dynamic range of the genre will still be heard. For pop music, more compression must be heard to ensure that little dynamic range is present and every sound is heard above all outside distractions.

12.7 Mix-Bus Processing

There are two schools of thought when it comes to applying effects to the master fader in a mix. Some engineers believe that no processing should be applied on the master, as issues and enhancements can be done on individual tracks. This way of thinking also extends to the fact that applying effects on the master is essentially mastering, although this statement is true only on the basis of when these effects are applied: it is still considered mixing if they are used at the start of the mix, and it is mastering if they are added at the end. Using effects on the master at the start of the mix can help the engineer do the "heavy lifting" once and therefore process less during the mix. For example, applying a broad top-end boost to the master can ensure that less individual track brightening is done. Another example is using gentle compression on the master to help glue instruments together instead of using multiple stages of glue compression throughout the mix. If you are using those processors at the start of the mix, it is important to follow two simple rules. First, you should use these processors as soon as you have recorded all the instruments and cleaned them up (strip silence, gain staging, filters, correlation, rough balance), but before you make any creative-mixing decision. The second rule is to get in a "mini-mastering"-session state of mind, but ensuring that the effects are used only for enhancement rather than fixing. This approach will ensure that you apply only gentle "broad strokes" to the mix. Overall, there is no right or wrong when using mix-bus processing, and this choice often comes down to the engineer's preference.

12.8 Hardware Emulations

New plug-in emulations of hardware units are being released every year. While some sound better than others, they all share one fundamental difference with their analog counterparts. When pushed hard, plug-ins start to "break down" a lot faster than hardware units. Large amounts of gain reduction on hardware compressors or large amount of boosts on hardware EQs can often be achieved without issues. When you are using plug-in emulations, the sonic characteristics of the original unit are often well represented, but they cannot be pushed as hard.

12.9 1001 Plug-ins

Do not download thousands of plug-ins with the thought that your mixes will sound better with them. In the first few years of being an audio engineer, you will have better-sounding mixes if you use one EQ and one compressor. Learn the stock plug-ins of your chosen DAW like the back of your hand and know their advantages and limits. When you finally start using higher-quality plug-ins, you will hear the differences right away and you will be able to judge which plug-ins are better than others for certain tasks. Once again, good mixing engineers can achieve great results by using only the stock plug-ins of their DAW; third-party plug-ins and other fancy tools are only making the mixing task easier.

12.10 Tactile Control

Tactile controllers can be very useful in achieving better results when mixing. It is a fact that when we use the mouse to tweak plug-in settings, we are relying on both ears and eyes to judge, for instance, the amount of EQ boost or compression. Working this way can be detrimental to the mix, as our eyes may be fooling us into thinking that there is too much (or too little) EQ, compression, and reverb, for example. Using MIDI controllers that can easily map plug-in parameters will help in using the only sense that should matter when mixing: your hearing.

12.11 Hot Corners

Whenever possible, try to mix without looking at the arrangement, to redirect your focus on what is being heard rather than seen. Both Macs and PCs have "hot corners" available where the mouse can be placed to engage a function that will hide the arrangement window. Using this technique has the benefit of revealing sections that work (or not) more easily. The expectation built up by seeing when the next section of the song comes in often blurs engineers' judgment on the effectiveness of a part.

12.12 When a Part Will Not Fit

Every now and then, an instrument or sound will not fit within the context of the mix and be very frustrating, as you can spend hours in trying to fit everything in without success. If something does not fit, the first question you should ask is whether the part is necessary in the mix. While the producer should answer this question, as the creative decision to remove a part should not be made by the engineer, sometimes there is no other choice but to mute an instrument for parts of the song. If the instrument must remain, a trick to try to fit it within the mix is to pull its fader all the way down and imagine how it should sound within the mix. What sort of EQ curve should it have? How far forward in the mix should it be? How dynamic should it be? How do its composition and rhythm work with the other instruments in the mix? Be sure to leave the instrument muted long enough to build a very good mental picture of it. Twenty to thirty seconds should be enough for this exercise. Once you have made a mental image of the instrument, raise its volume and compare the differences with the sound imagined. It is important to act on those differences immediately. By using EQs, compression, reverb, volume, panning, and so on, you should be able to turn the offending part into something that fits within the mix.

Chapter **13**

Panning

13.1 Intensity versus Phase Stereo

Humans have two ears and therefore hear sounds in stereo. There are two ways that we perceive the direction of a sound in the stereo field: through differences in intensity (both general and frequency specific) and arrival-time differences.

AUDIO PRODUCTION PRINCIPLES

These two ways of perceiving the direction of a sound can be used by the mixing engineer to place instruments within the stereo field.

Intensity stereo works by varying the amplitude of a signal between the two speakers (Figure 13.1; Audio 13.1 ▶). If the same signal is played more loudly in the left speaker than it is in the right speaker, we will perceive it as coming from the left. Since pan pots are used to achieve this effect, it is the easiest way of placing sounds in the stereo field.

FIGURE 13.1
Intensity stereo.

Spectral differences between sounds in the left and right channels can also be used to place instruments (Figure 13.2; Audio 13.2 ▶). Because our skull effectively blocks high frequencies, hearing the same sound in both ears with one side being brighter than the other will give the illusion of it coming from the bright side. While this perception is technically a form of intensity stereo (frequency-specific amplitude difference), dual-mono equalizers can be used to place elements toward one side or the other.

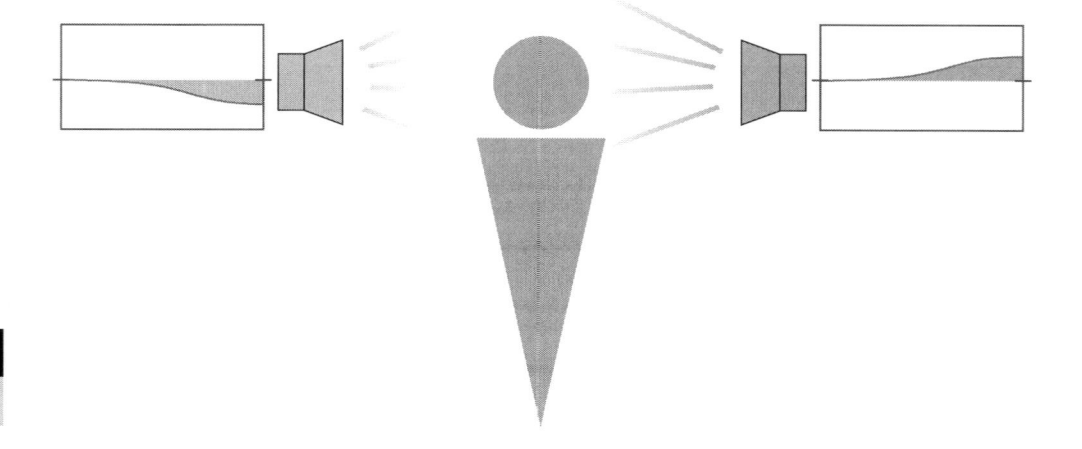

FIGURE 13.2
Spectral-intensity stereo.

Phase stereo is achieved through differences in the time it takes a sound to reach one ear or the other (Figure 13.3; Audio 13.3). If a sound is played in the left speaker slightly after it is played in the right speaker, we will perceive it as coming from the right. This technique works best with delays ranging from 3 ms to 30 ms.

FIGURE 13.3

Phase stereo.

Using this technique to push an instrument to one side or the other often sounds more natural than using the pan pot. While short delays are used to achieve this effect, you should be careful, as phasing issues can arise if you improperly set the delay. Always check the mono compatibility after panning an instrument by using this technique. Using phase stereo to place sounds to one side or the other also tends to push them back in the mix (Figure 13.4). This placement can be of great help in dense mixes in which a new area of the virtual soundstage can be filled.

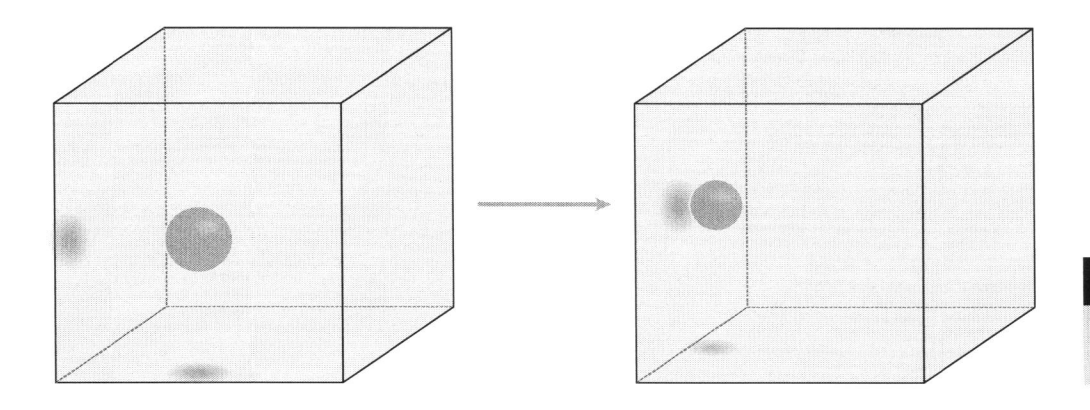

FIGURE 13.4

Effect of phase stereo on depth position.

When you are using phase stereo to push a sound to one side and using intensity stereo to bring it back to the center, a sound will get the blur from being further back in the mix while remaining at the same volume level. Roughly

10 dB of boost on the opposite side is necessary to re-center the signal after panning by using phase stereo (Audio 13.4 ▶).

While both intensity and phase stereo can be used during mixing, intensity stereo is more effective for frequencies above 4 kHz, and phase stereo works better for frequencies below 2 kHz.

13.2 Pan Law

The pan law dictates the volume of an audio signal as it is panned across the stereo field. Modern DAWs offer various pan laws such as 0 dB (Figure 13.5), –3 dB (Figure 13.6), –4.5 dB, and –6 dB, compensated or not. As a signal is panned from the center to one side, it will gradually drop in perceived volume, as only one speaker will play it back. This effect is known as the 0 dB pan law, in which the volume stays at its nominal level. In order to keep a perceived constant volume while panning across the stereo field, you must use other pan laws. The –2.5 dB, –3 dB, –4.5 dB, and –6dB pan laws effectively turn the gain of the signal down by the chosen value (–2.5 dB, –3 dB, –4.5 dB, and –6 dB) when the signal is panned center. This adjustment gives the illusion of constant volume when you are panning across the stereo field. In a rule of thumb, the better the acoustics in the mixing room, the lower the pan law that can be used. The level of –2.5 dB was the original pan law used in Pro Tools. The level of –3 dB is generally accepted as being the norm for most program material and mixing rooms. The level of –4.5 dB is the pan law used in SSL mixing consoles. The level of –6 dB is used for testing purposes or in extremely well-treated rooms that would give the true representation of how perfectly in-phase signals may combine to give a perceived 6 dB boost in volume. Some pan laws are also compensated (–3 dB compensated, for example, as is the case in Logic Pro and some Yamaha digital consoles). This compensation is very similar to the non-compensated pan laws, except that instead of turning the center down, you raise the sides by the chosen value.

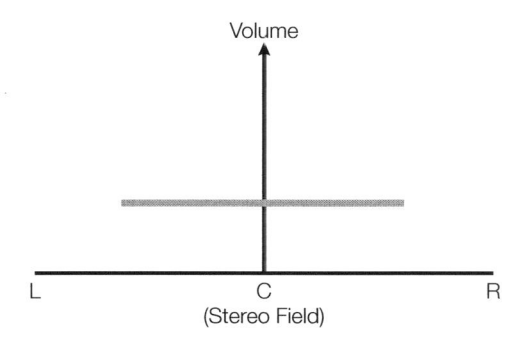

FIGURE 13.5

Zero dB pan law.

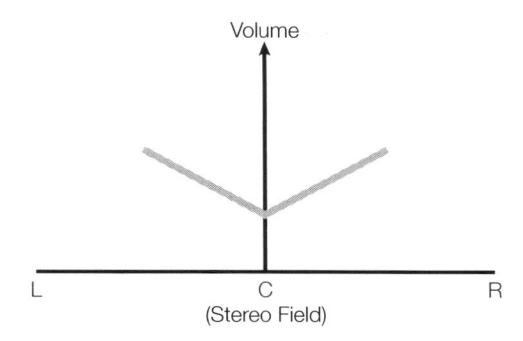

FIGURE 13.6

Minus 3 dB pan law.

13.3 Panning Tips

13.3.1 Multi-Mono Equalizers

While the concept of using EQs to make one side brighter than the other in order to pan an instrument in the stereo field has already been explained, we can push this technique further. Using different EQ shapes on each side of stereo instruments will enlarge the instruments without the need to pan them further to the sides (Figure 13.7; Audio 13.5 ▶). Try equalizing the left and right channels of the drum bus differently to illustrate this technique.

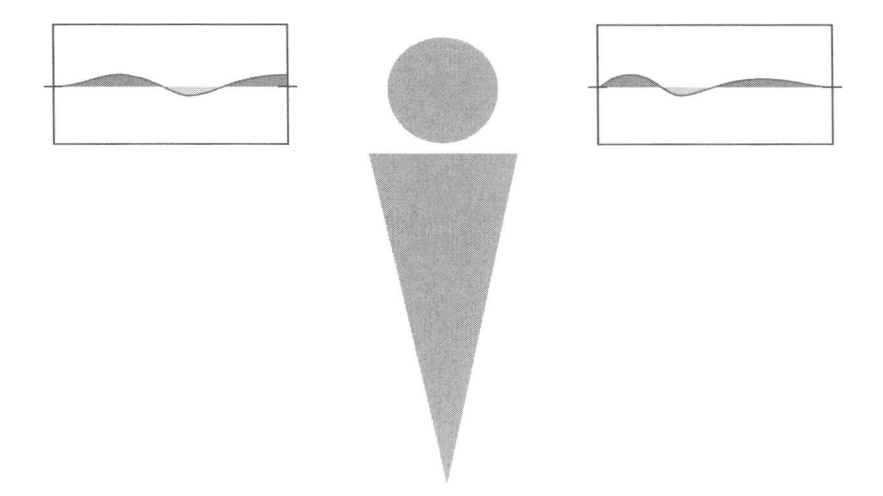

FIGURE 13.7

Decorrelation using multi-mono EQ.

This concept can also be used to pan instruments by frequency. Think of multi-mono EQs as a way of panning frequencies within instruments rather than the whole sound itself—for example, thinking of a recording containing only vocals, acoustic guitar, and banjo. In order to create separation between the guitar and the banjo, you might choose to pan them on opposite sides, with the voice left in the center. The issue with this method is that the low end given by the acoustic guitar would then be fully on one side, unbalancing the mix. The answer to this problem would be to use a multi-mono EQ on the guitar by using high shelves to push the mids and highs to one side (by boosting them on one, and cutting them on the other), while leaving the low end intact.

This step would effectively keep the bass present in the guitar in the center of the mix while you pan the mids and highs to one side to counter balance the banjo on the other.

13.3.2 Mono to Stereo

There are a few tricks to turn a mono signal into stereo. While there is no better solution than recording in stereo, some tricks can be used to create the illusion of a stereo recording. Duplicating the mono track and panning hard left and right is the first step. The following processors can then be used to create the illusion of stereo:

- Pitch-shifting one side a few cents up or down (Audio 13.6 ▶). Using a doubler FX on one side also achieves the same result in a non-static way.
- Adding a short delay to one side (Audio 13.7 ▶). Using flanger FX on one side achieves the same result in a non-static way.
- Using a phaser FX on one or both sides (Audio 13.8 ▶).
- Using different compression settings on either side can also create movements between the speakers (Audio 13.9 ▶).

Care should be taken, as the techniques involving time delays and phase shifts will affect mono compatibility as phasing between one channel and the other is introduced.

One way to create a fake stereo signal while ensuring that the mono information remains intact is to keep the original signal in the center, duplicating it twice (panned hard left and right), time-delaying the duplicates by the same amount, and reversing the polarity of one (Audio 13.10 ▶). Using this technique means that the comb filtering created by delaying the doubles will be canceled in mono, as the duplicates are 180 degrees out of phase with each other.

Instead of duplicating the original track, you can also pan it to one side and send it to a mono reverb on the other (Audio 13.11 ▶). While this method also pushes the sound to the dry side, it is a fairly natural way of adding width to a mono signal.

13.3.3 Mid-Side versus Left-Right

In order to enhance the separation between instruments, using MS processors such as an MS EQs rather than stereo versions can give interesting results. For example, the separation can be achieved by removing an instrument from the phantom center or shaping its tone, depending on its placement and width. The different ways to encode and decode a signal to access the middle or the sides separately are described in more detail in Chapter 18, "Mastering." If we take this concept further, it is possible to compress the sides of a signal to ensure that the stereo width is constant rather than dynamic (i.e., the width changing, depending on the dynamics of the instruments panned to the sides). It is also possible to use this technique in parallel to raise the low-level signal on the sides rather than lower the high-level signal.

13.3.4 Mid-Side Equalizer Width

Using a low shelf to boost bass and low mid-frequencies on the sides can add width to a stereo signal while it remains fully mono compatible (Audio 13.12 ▶). Adding an HPF to ensure that no sub-bass was added to the sides is often useful as well. This technique can also work with a harmonic enhancer placed on the low end instead of using a low shelf.

13.3.5 Hard Panning

Avoid panning a mono instrument completely to one side. Hard-panning an instrument to one side means only that it will completely disappear if one speaker channel gets dumped. This result is often the case in nightclub systems, TVs, and some boom boxes.

13.3.6 Wider Mixes

Given that the stereo field is one full dimension of the virtual sound-stage, achieving wide mixes (that are still mono compatible) is important in making the mix fuller and more interesting. Just as a surround-sound mix brings the listener "into" the virtual performance stage, a wide stereo mix can pull the audience in rather than make it listen to the song from a distance.

In order to achieve wider mixes, turn more instruments mono and spread them across the stereo field. If every stereo instrument is panned hard left and right, the result will have something of a "wide mono" image (Figure 13.8). Extra width cannot be achieved this way. Panning instruments solely to one side can give the illusion that the mix is much wider by introducing noticeable differences between the speakers (Figure 13.9; Audio 13.13 ▶). When you are using this technique, it is important to remember to always balance the stereo field with instruments of similar frequency content on either side. A hi-hat panned on the left should be counterbalanced with a triangle or ride cymbal on the right, for example.

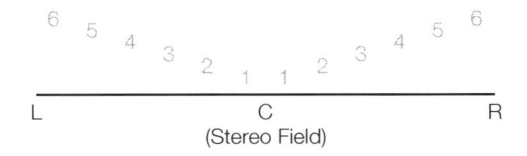

FIGURE 13.8
Narrow mix.

FIGURE 13.9
Wide mix.

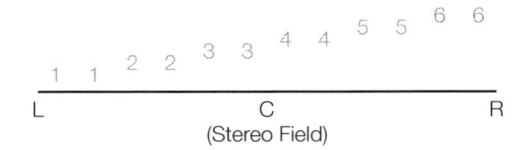

13.3.7 Correlation Meters

Always check the correlation meter when you are altering the stereo width of a signal (through enhancers, delays, and MS processors, for example). While a wide stereo image is almost always desirable, more importance should be placed on the mono compatibility of the mix. In a rule of thumb, staying around the halfway mark on the positive side of the meter provides a good balance between stereo width and mono compatibility.

13.3.8 Backing Vocals

While lead vocals are almost always placed in the center of the stereo field, backing vocals are often spread more widely to increase their perceived size. Since most instruments containing low-end frequencies will end up in the center, using the opposite approach when you are placing backing vocals works best. Low harmonies should be placed further to the side than high harmonies to provide the widest stereo image for backing vocals.

13.3.9 Panning Instruments

Instruments such as synth pads and distorted guitars playing chords, and others that do not have clear transients are often double tracked (if not triple or quadruple tracked!). This technique is sometimes referred to as the "wall of sound." For instruments that have been recorded in stereo, there are different choices available when it comes to panning. In the case of a recording that has been done with two microphones (M1 and M2) and doubled once (M1´ and M2´), two main choices are available (Audio 13.14 ▶):

- Left: M1 + M2
- Right: M1´ + M2´

This method achieves more separation and clarity (Figure 13.10).

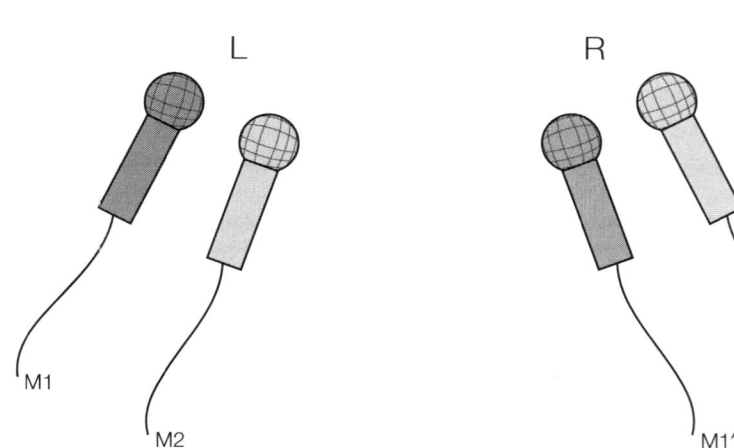

FIGURE 13.10
Panning double-tracked instrument: separation and clarity.

- Left: M1 + M1´
- Right: M2 + M2´

This achieves more power and fullness (Figure 13.11).

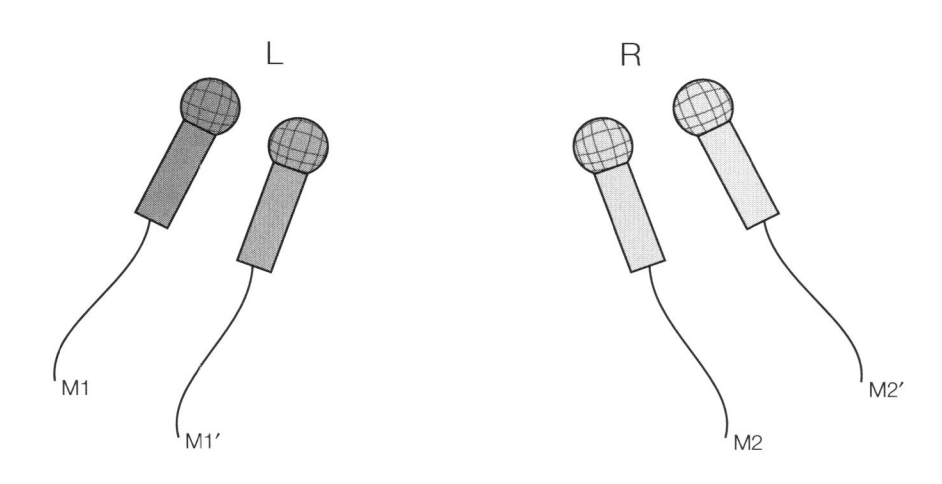

FIGURE 13.11
Panning double-tracked instrument: power and fullness.

13.3.10 Panning for Depth

In the physical world, when we hear sounds from a distance, they appear to be mono. The stereo aspect of an instrument is perceived only as it is brought closer to the listener. You can use this phenomenon when you are mixing, as making an instrument mono can push it back in the mix (Figure 13.12), while making it wider can bring it forward (Figure 13.13).

FIGURE 13.12
Panning for depth: wider instrument feels closer to the listener.

AUDIO PRODUCTION PRINCIPLES

FIGURE 13.13
Panning for depth: narrower instrument feels farther away from the listener.

Chapter **14**

Equalizers

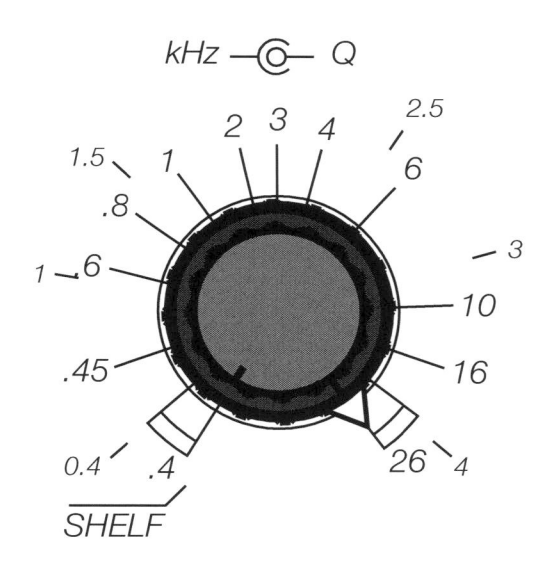

Equalizers are one of the most important tools used for mixing. While general equalizing guidelines have been explained in Chapter 12, "Mixing," extra information is necessary in order to fully utilize these tools. The secret to using EQs properly is being in the right "mindset," and equalizing for a reason. It can be useful to think of EQs as frequency-dependent volume controls. After all, these tools are used only to raise or lower the volume of individual frequencies, similar to what an overall volume control is, but with more precision. There are three different uses for EQs: fixing, enhancing, and separating instruments. The type of EQ and the filter type, gain, and bandwidth (or Q) settings will all be dictated by the reason behind equalizing. Make sure that you know why the EQ is being used before patching anything, as this knowledge is the only way to get the best possible sound.

14.1 Fixing Equalizers

Fixing EQs are used for removing frequency content from instruments or when needing to transparently increase the volume of a particular frequency. A good rule of thumb to follow is to never boost more than 4–5 dB. If more boost is necessary, either the overall volume of the instrument is too low, or an enhancement EQ may work better. In much the same way, if more than 10 dB of cut is needed, the filter probably is the wrong type. In this case, changing a low shelf to a high pass, a high shelf to a low pass, and a bell to a notch filter may work best. The Q setting for both boosts and cuts should be quite narrow to ensure that the EQ targets specific problem frequencies (Figure 14.1). Clean digital EQs such as the stock equalizers that come bundled with DAWs work well for this purpose. Fixing EQ should be done as early as possible in the project. When fixing instruments with EQs, ensure that the original tonal qualities of the instrument are kept intact. The mindset to be in when fixing is to remain transparent at all times! Note that this type of equalizing is generally done with instruments soloed.

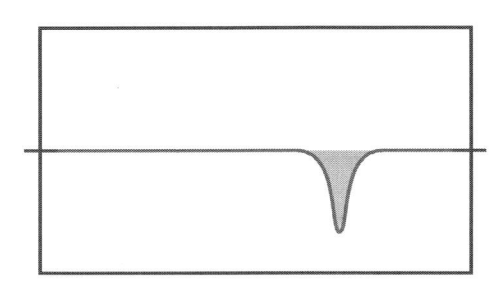

FIGURE 14.1
Narrow EQ cut for fixing recorded tone.

14.2 Separation Equalizers

Separation is the most important use for EQs, as this process can make or break a mix. While this topic has already been discussed in Chapter 12, it is important to include it here to differentiate equalizing for separation from the other two types of mindsets possible. The mindset to be in when you are separating is to give each instrument as much space as possible without any sound being overpowered by others or sounding too "small." Note that this type of equalizing is generally done with groups of instruments of the same frequency content soloed (for example, kick and bass, bass and guitars, and guitars and vocals).

14.3 Enhancement Equalizers

These processors are often used when large amounts of boosting are necessary or when the tonal quality of the EQ is to be imprinted on the signal. More "exotic" equalizers are necessary for this purpose. Neve, SSL, and Pultec EQs are examples of processors that have their own sonic character, which is often sought after. It is not uncommon to use clean EQs to remove frequency content from a track, only to bring that same frequency back with a character EQ. Enhancement EQ is more often used at the mixing stage of a project to make instruments sound better than they originally did on the recording. To this end, wide Q settings are mostly used (Figure 14.2). The mindset to be in when enhancing is to make instruments sound as best as they can without overpowering others. Note that this type of equalizing is generally done with every instrument playing.

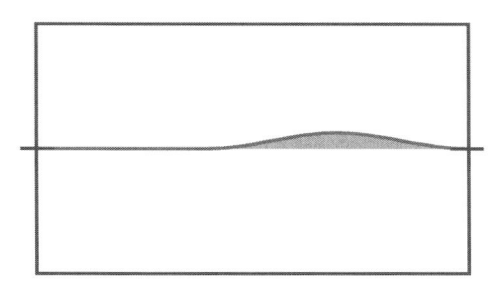

FIGURE 14.2 Wide EQ boost for enhancing recorded tone.

14.4 Q Setting

Choosing a wide or narrow Q (Figure 14.3) depends largely on the task at hand. A narrow Q is often used to remove unwanted ringing frequencies (notch filters are often used for this task) or to add a very specific sound in low frequencies, such as the "thump" or "boom" in a kick. Wider Q settings are used for broader and smoother boosts or cuts (Box 14.1). Generally, narrow boosts are more noticeable than wide ones. Narrow and wide cuts are similar in terms of how noticeable they are.

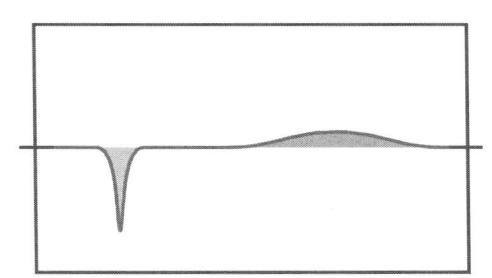

FIGURE 14.3 Different Q settings: narrow cut and wide boost.

BOX 14.1

Some instruments such as cymbals and distorted guitars can have constant ringing in their sound. While notch EQs are excellent at resolving these issues, it is important not to look for them by using the "boost/sweep/cut" technique. The ringing in a cymbal is an issue only if it can be heard without the need for a narrow boost and sweep in the high end (Audio B14.1).

FIGURE 14.4
Gain-Q interaction: Q setting narrows as gain increases.

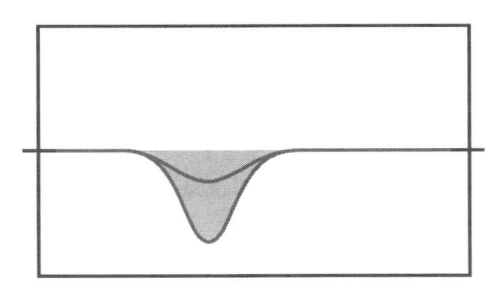

Some EQs have gain-Q interactions where the Q tightens as gain increases. This is a useful way of "intelligently" equalizing, where large cuts often need to be quite narrow, and small boosts are wide (Figure 14.4).

Boosts and cuts in EQs can either be symmetric (Figure 14.5) or asymmetric (Figure 14.6). In symmetric EQs, the boosts and cuts are mirrored at the same positive and negative gain settings. In asymmetric EQs, cuts are narrower than boosts. This mirroring replicates the way of working described with gain-Q interactions where boosts usually need to be wider than cuts to remain natural.

FIGURE 14.5
Symmetric Q settings.

FIGURE 14.6
Asymmetric Q settings.

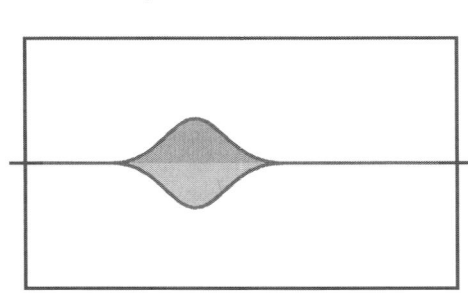

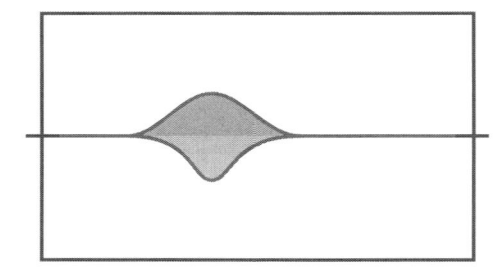

14.5 Filter Type

HPFs and LPFs can be designed in a number of ways (Audio 14.1 ▶). While an in-depth knowledge of filter design is not necessary in order to mix music, it is useful to understand that different filters are best suited for different uses. The most common filter designs are the following:

Butterworth filters (Figure 14.7) are smooth with no ripples in the frequency content left after filtering.

FIGURE 14.7
Butterworth LPF.

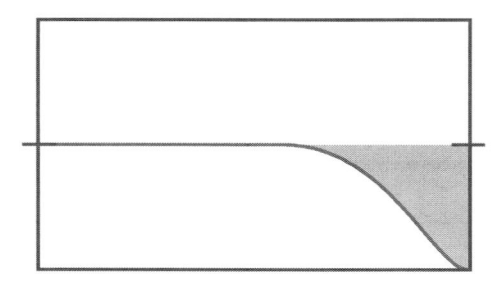

Legendre filters (Figure 14.8) can go more steeply than Butterworth, while still remaining smooth in their frequency response.

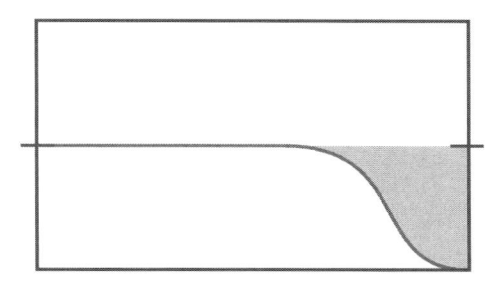

FIGURE 14.8 Legendre LPF.

Chebyshev filters can attenuate more frequencies, but at the expense of some ripples in the passband or stop band (for type 1 [Figure 14.9] and type 2 [Figure 14.10] respectively).

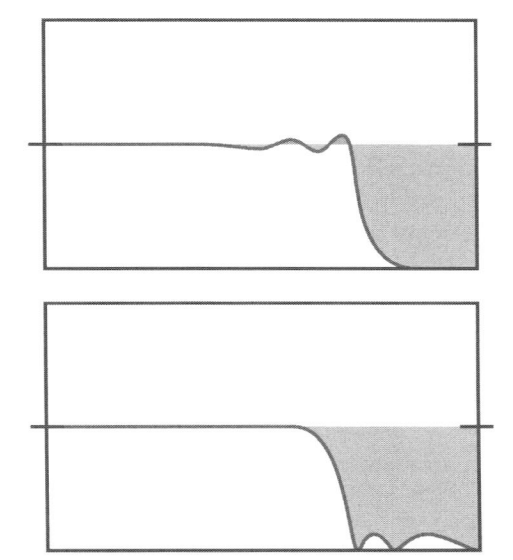

FIGURE 14.9 Chebyshev type 1 LPF.

FIGURE 14.10 Chebyshev type 2 LPF.

Elliptic filters (Figure 14.11) can attenuate even more, but contain a lot of ripples.

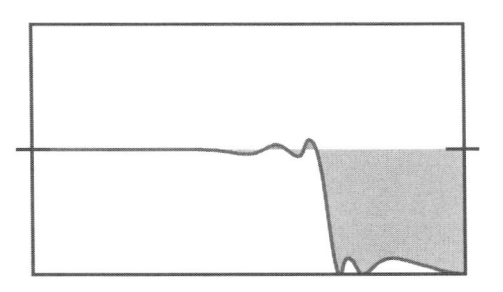

FIGURE 14.11 Elliptic LPF.

14.6 Linear-Phase Equalizers

All analog EQ filters introduce phase shifts in the signal being processed. Whether boosting or cutting, or using bells, shelves, or HPFs and LPFs, there is always some amount of phase shift in the frequencies that remain. The phase-shift shape depends on the type of filter being used (bell, shelf, notch, HPF, and LPF all have different phase shifts, with HPFs and LPFs yielding the most phase shifts), the amount of boost or cut being applied (the bigger the boost or cut, the more phase shift), and the Q setting (the steeper the Q, the more phase shift). This phase shift is sometimes desirable and is the reason that hardware units have a particular sound. It is important to note that this phase displacement should be carefully considered when you are mixing two signals of the same frequency content (low frequencies are particularly affected by this displacement). For example, when high-pass-filtering one snare microphone, and leaving the second one unfiltered, the fundamental frequency of the snare may be out of phase between one microphone and the other. Another example would be to equalize in parallel by duplicating a track and filtering only one copy. In this case, you can achieve interesting results, as the unfiltered frequencies may be phase displaced and therefore attenuated or boosted. Since the phase-shift shape can vary greatly between EQs models and settings, it is important to listen for the changes in perceived frequency content even in the unfiltered parts of the signal.

At times, transparent equalization is necessary, which is what linear-phase EQs are designed for. In the digital domain, it is possible to avoid the phase shifts by using such EQ. The result, although sometimes referred to as being "cold" or "digital" sounding, is very transparent and predictable. Such EQs are often used in mastering to remove problem frequencies without affecting the tonal quality of the original mix. All the previously mentioned considerations for mixing signals of similar frequencies are unnecessary when you are using linear-phase EQs, as there is no phase displacement involved (Figures 14.12 and 14.13; Audio 14.2 ⏵). Note that linear-phase EQs require a lot more processing power than regular digital EQs and thus are impractical for mixing. This impracticality is because they are implemented by using a finite impulse response. It is possible to approximate the same behavior by using an infinite impulse response, which is less CPU intensive, through Bessel-type filters. Another drawback of linear-phase EQs is the potential for pre-ringing, which can become quite apparent in low frequencies. Pre-ringing happens because of the overall delay caused by the EQ being compensated by the DAW. Note that the steeper/sharper the Q, the more pre-ringing becomes apparent. This pre-ringing can become an issue when you are processing sounds that contain clearly defined transient and sustained parts, such as a kick or

FIGURE 14.12

Phase distortion in linear-phase EQ.

FIGURE 14.13

Phase distortion in minimum-phase EQ.

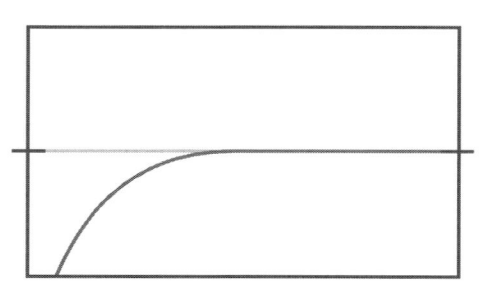

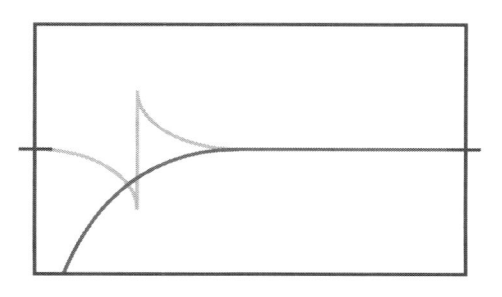

snare drum. In such a scenario, there is a potential for the pre-ringing to blur the initial transient.

14.7 Active and Passive Equalizers

Put simply, active EQs differ from their passive counterparts by having circuitry that allows them to both boost and cut on a band-by-band basis (Figure 14.14). Passive EQs, on the other hand, work only with cuts, not boosts (Figure 14.15). An amplifier is then placed at the output of the device to raise the overall volume, if necessary. If a boost is dialed in on such EQ, the frequency chosen is merely "less attenuated" than others, and then the overall output level is raised. Passive EQs are often used for their character as enhancement devices (Audio 14.3 ▶).

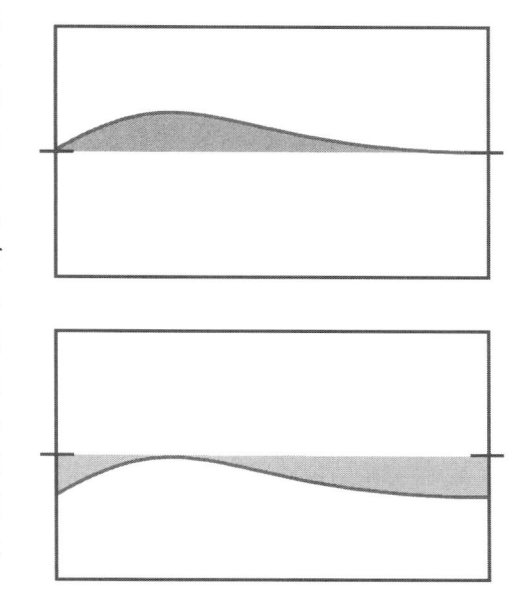

FIGURE 14.14
Low-frequency boost made with active EQ.

FIGURE 14.15
Low-frequency boost made with passive EQ.

14.8 Equalizer Tips

14.8.1 Objective versus Subjective Descriptive Terms

Instrument sounds are often described by using subjective terms rather than their frequency content, as it is easier for the mind to imagine how an instrument sounds by using those terms. It is therefore important to understand what frequencies constitute each instrument, and which of them are being associated with common subjective terms. Figure 14.16 lists the main frequency content of the instruments mentioned in this book as well as specific terms used when a lot of energy is present at particular frequencies.

Instrument Frequency Content

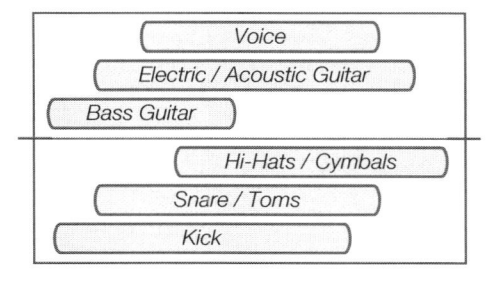

- Voice
- Electric / Acoustic Guitar
- Bass Guitar
- Hi-Hats / Cymbals
- Snare / Toms
- Kick

Subjective Frequency Terms

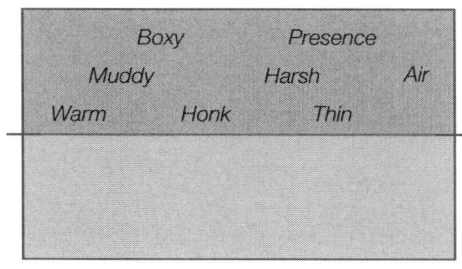

Boxy   Presence
Muddy   Harsh   Air
Warm   Honk   Thin

FIGURE 14.16
Instrument frequency content and subjective frequency terms.

14.8.2 Cut Rather Than Boost

With the exception of high-end EQs, and in very specific applications (read "enhancement EQ"), it is always best to cut instead of boost. This guideline should be applied for the following two reasons:

- Most equalizers deteriorate sounds when boosting frequencies. Or more precisely, sounds attract attention to themselves when particular frequencies are boosted a lot (especially with narrow Q), making the change more obvious and unnatural. As previously mentioned, high-end equalizers are used for boosting because of the tone and character they bring. General-application EQs such as those used for fixing and separation are best used for cutting only.
- The mindset of "cutting only" can force you to think more critically and creatively about the equalizing process. When limiting yourself to cuts rather than boosts, you must employ different approaches that often yield better results than going with the easy solution that is boosting. For example, in order to bring out the detail in the electric guitars, you could boost 6–8 kHz. If you are thinking in terms of cuts only, reducing 300–500 Hz on the guitars and 6–8 kHz on the piano may give the same overall effect, but in a much cleaner way. Think of every EQ move you do as having a "C-saw" effect on the frequency content: if you cut the lows, it will sound similar to boosting the highs. This approach means that in order to make an instrument brighter, for example, boosting its high frequency is not the only EQ move possible. Cutting its low end could work just as well, if not better. It is important to note that this C-saw effect happens on the main frequency content of the instrument being equalized. For example, cutting 300 Hz on a voice would make it sound thinner because its frequency content mostly ranges from 200 Hz to 8 kHz, whereas cutting 300 Hz on a bass would actually make it sound fuller because its frequency content mostly ranges from 50 Hz to 500 Hz (Figure 14.17; Audio 14.4 ▶).

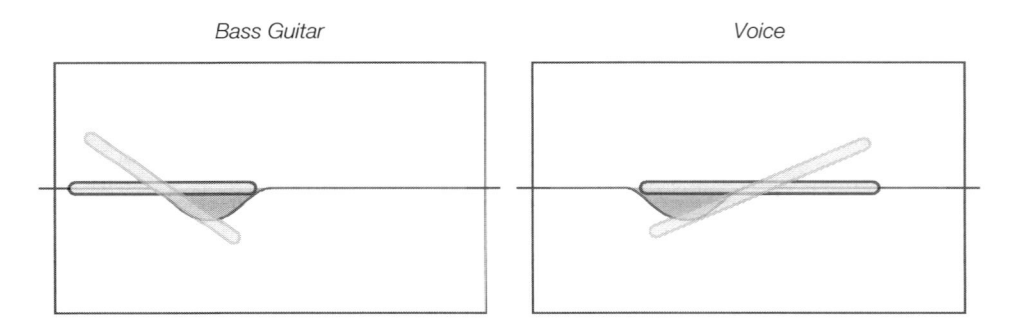

FIGURE 14.17

"C-saw" effect of EQ on bass versus voice.

14.8.3 Keeping in Tune

While engineers easily understand the need to tune guitars and other instruments, they often overlook other aspects of this important process. Working out the key of the song early will help in making the right EQ moves during mixing. Ensuring that every EQ boost made follows the harmonic series (Figure 14.18) of the root note in the key of the song often means that any boost works "with" the music.

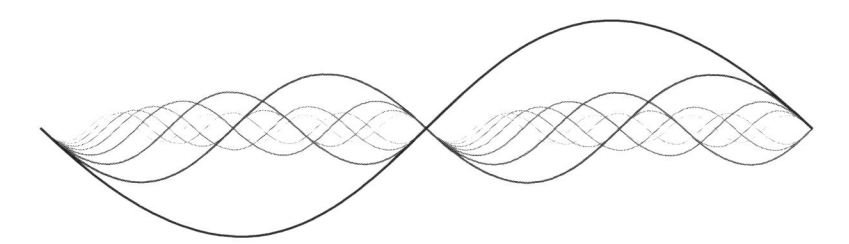

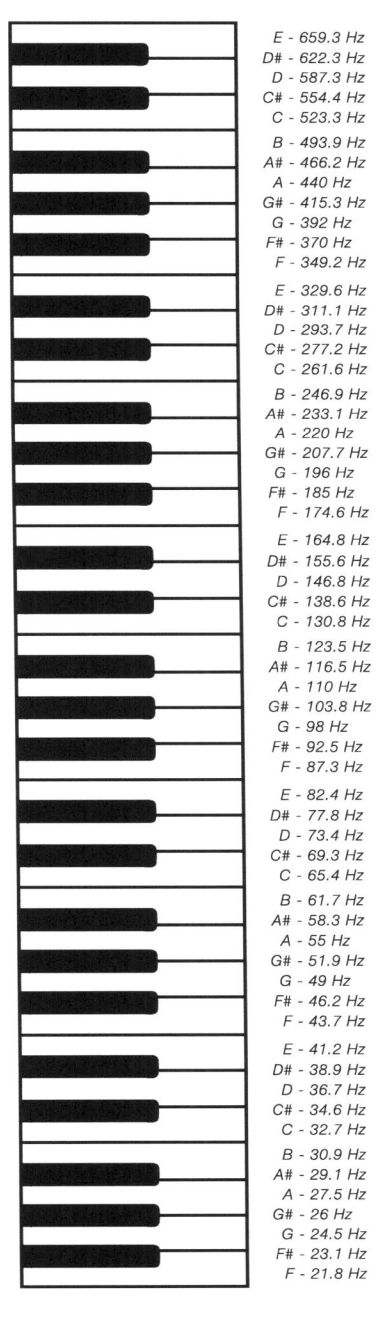

FIGURE 14.18

Visual representation of the harmonic series.

Generally speaking, it is a good idea to keep boosts, especially in the low end, to the root note and fifth to ensure you are keeping in tune. Use Figure 14.19 as a guide to select frequencies to boost in the low end, keeping in mind that this technique does not usually work for melodic instruments, as boosting a specific note may alter the performance's volume level. Note that for bass guitars, this boost may result in particular notes being louder than others. You will then need to bring these notes back in volume, or alternatively, use wider Q settings to ensure that no one note is emphasized.

14.8.4 Equalizing for Depth

In the physical world, as a sound source is getting farther away from the listener, it loses high-frequency content. Dull sounds feel farther away than bright sounds. This phenomenon can be used to push instruments back in the mix by using a high shelf to lower the high-frequency content (Audio 14.5 ▶). This method is especially effective when followed by reverb, which will push the instrument further back in the mix.

E - 659.3 Hz
D# - 622.3 Hz
D - 587.3 Hz
C# - 554.4 Hz
C - 523.3 Hz
B - 493.9 Hz
A# - 466.2 Hz
A - 440 Hz
G# - 415.3 Hz
G - 392 Hz
F# - 370 Hz
F - 349.2 Hz
E - 329.6 Hz
D# - 311.1 Hz
D - 293.7 Hz
C# - 277.2 Hz
C - 261.6 Hz
B - 246.9 Hz
A# - 233.1 Hz
A - 220 Hz
G# - 207.7 Hz
G - 196 Hz
F# - 185 Hz
F - 174.6 Hz
E - 164.8 Hz
D# - 155.6 Hz
D - 146.8 Hz
C# - 138.6 Hz
C - 130.8 Hz
B - 123.5 Hz
A# - 116.5 Hz
A - 110 Hz
G# - 103.8 Hz
G - 98 Hz
F# - 92.5 Hz
F - 87.3 Hz
E - 82.4 Hz
D# - 77.8 Hz
D - 73.4 Hz
C# - 69.3 Hz
C - 65.4 Hz
B - 61.7 Hz
A# - 58.3 Hz
A - 55 Hz
G# - 51.9 Hz
G - 49 Hz
F# - 46.2 Hz
F - 43.7 Hz
E - 41.2 Hz
D# - 38.9 Hz
D - 36.7 Hz
C# - 34.6 Hz
C - 32.7 Hz
B - 30.9 Hz
A# - 29.1 Hz
A - 27.5 Hz
G# - 26 Hz
G - 24.5 Hz
F# - 23.1 Hz
F - 21.8 Hz

FIGURE 14.19

Note-versus-frequency chart.

14.8.5 Equalizer Matching

Some equalizers allow the engineer to analyze a sound and match its overall tone, or create the exact opposite, with another. While this sort of processing is not used very often, it has some advantages over conventional equalizing. For beginner engineers, it is a good way to see what sort of EQ curves are needed in order to match a sound to its reference material. It is also quite useful for quickly making two sounds work around each other by accentuating their differences by using opposite EQ curves. Lastly, it can help sounds recorded during different sessions blend better with one another by applying a similar overall EQ curve to them. This last use is probably the most important, as it could help ensure, for example, that two guitar recordings done on different days sound similar enough to be used in the same song.

14.8.6 Phase Trick

One way to figure out exactly what an EQ is adding or removing is to duplicate the track, flip the phase on one copy, and EQ it. Because of the phase inversion, only boosts or cuts made with the EQ will be heard. This trick can be a great way of figuring out if an EQ sounds harsh when you are boosting high frequencies or the tonal quality of midrange boosts, for example. This technique can fast-track the learning of different EQ plug-ins, as you can dial in similar settings on different plug-ins, then audition the settings one at a time to judge how each EQ sounds.

14.8.7 Low-Mid Frequencies

Quite often, instruments benefit from having less power around the 300Hz–600 Hz area. This frequency area tends to emphasize "boxiness" and bring out "mud." The issue with removing these frequencies from everything in the mix, alongside using HPF for instruments that do not extend to the bass range, is that the mix overall may feel thin. It is important that some instruments retain those frequencies to add weight to the mix. The trick is often to remove a lot of low mids from everything except for a few instruments placed at the front of the mix. While background instruments do not need these frequencies, foreground elements can carry the weight of the whole mix by retaining energy in the low mids.

14.8.8 High Frequencies

Some equalizers allow for boosts above 20 kHz. Such EQs can generally add "air" and brighten sounds without adding harshness sometimes present when you are boosting 10–20 kHz (Audio 14.6 ▶). Even though the frequencies added are above our hearing range, the effect they have on the signal below 20 kHz can be quite pleasing to the ear. The higher the frequency added, the more its effect extends toward the low end of the spectrum. This factor means that a boost at 40 kHz will extend lower in the spectrum than a boost at 26 kHz, for example. Baxandall high shelves (Figure 14.20) can achieve similar results, as they have very wide Q and therefore keep rising above 20 kHz instead of flattening as a normal shelf would.

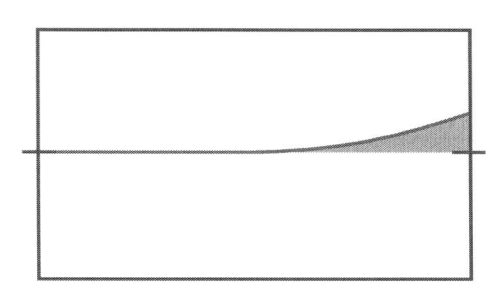

FIGURE 14.20
Baxandall high-shelf EQ curve.

14.8.9 Side-Chained Dynamic Equalizer

In an effort to achieve maximum separation between instruments, engineers sometimes use side-chain compression turn the volume of one instrument down while another is playing. When we push this concept one step further, it is possible to equalize some frequencies out of an instrument dynamically (Figure 14.21) (rather than the whole signal) by using a multiband compressor or dynamic EQ with side-chain input capabilities (Audio 14.7). A common example of this method is to turn down the mids in guitars when vocals are present, and revert to a full-frequency guitar track when the singing is over. This step has the advantage of being subtler than regular side-chain compression, which should be used only for instruments of the same bandwidth, such as kick and bass. Some multiband compressors and dynamic EQs can also work in MS mode, which can help make this process even more transparent. For example, the lead vocals could be triggering a mids dip in the guitars in the center of the soundstage only. Dialing in a mids boost on the sides at the same time can help the guitars regain the presence lost from the dip in the center without masking the vocals panned center.

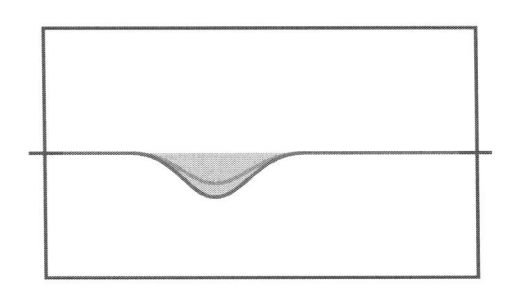

FIGURE 14.21
Gain moves dynamically, depending on side-chain input level.

14.8.10 Pultec Push-Pull Trick

Pultec EQs (Figure 14.22) can achieve resonant low shelves by boosting and cutting at the same frequency. The attenuation in such equalizing is roughly at twice the frequency of the boost, allowing for deep bass to be added to the signal.

FIGURE 14.22
Pultec EQ.

FIGURE 14.23
Gerzon low-shelf EQ curve.

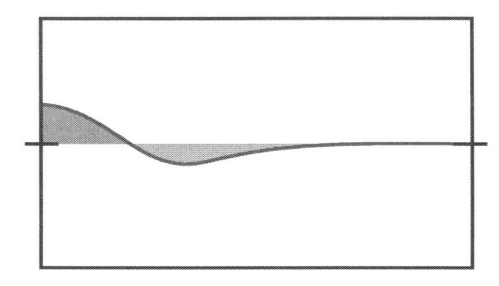

Engineers often use this technique as a parallel processor to add low-end content to kick drums without losing the original tone and punch of the kick. The Gerzon-type shelves (Figure 14.23; Audio 14.8) offered on some EQs can achieve similar shelves.

14.8.11 Equalizing Vocals

Boosts are often added at the frequencies where an instrument should "live" within the available spectrum. This technique does not work well for vocals, as pushing their mids tend to give very "nasal" or "harsh" results. Instead, ensuring that the mids are being cut in all other instruments is much more effective. Because we (human beings) hear people talk every day, we notice the sound of a voice that has been equalized in the midrange very easily. Slight brightening and low-end boosts can work fine, as we are less "tuned" to those frequencies (Audio 14.9).

14.8.12 Listening

When you are equalizing instruments to make them fit around one another, it is often preferable not to listen to the instrument being equalized, but instead to listen to its effect on others. For example, when you are fitting a guitar around vocals, the mids must often be cut in the guitar. In order to find the best frequency to cut, sweep the cut between 500 Hz and 3 kHz while you are listening to the effect it has on the vocals. You will find that with one particular frequency area cut on the guitar, the vocals come out of the mix and really shine through (Audio 14.10).

14.9 Plug-in Presets and Graphic Equalizers

Presets on EQ plug-ins should never be used. The reason is simple: how does the person programming the preset know which sounds you are starting from? For example, how does he or she know that the "fat snare" preset will not muddy the mix? Using EQ presets is similar to letting someone mix the song without being able to hear it. Sounds crazy, right?

The reason that amateur engineers use presets is simple: they want their instruments to sound better than what was recorded, but do not really know how or where to start. This is where graphic EQs come in handy. While graphic EQs are mostly used in live sound to tune rooms and get rid of feedback, they have their use in the mixing studio as the "preset" EQ. The concept is simple: if amateur engineers sift through presets until they find one that they like, why not boost every slider on a graphic EQ to audition the frequencies that they like (or dislike) instead? To that end, each band of the graphic EQ should be turned all the way up, one at a time, to hear what they sound like. If one band makes the instrument sound better, leave it up. Depending on how much "good" the boost is doing to the instrument, boost more or less of that particular frequency. If the boost makes the instrument sound worse, turn that frequency down. Once again, depending on how bad that frequency was, turn it down a little or a lot, thus effectively equalizing the instrument on the basis of what feels good, rather than what is "technically right" (Audio 14.11 ▶).

Chapter 15

Compression, Expansion, and Gates

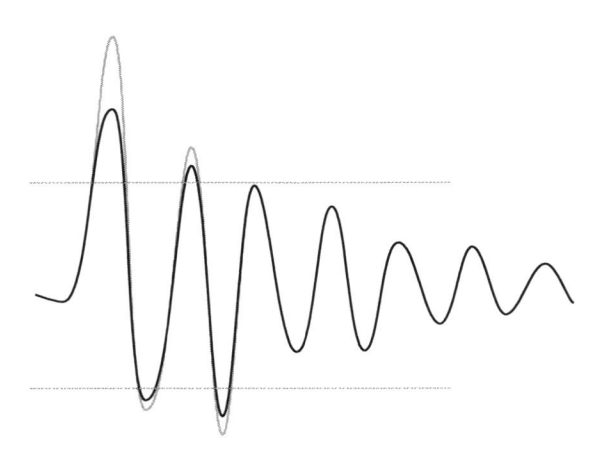

Compression is one of the most misunderstood tools used by audio engineers. It can be used to increase or decrease dynamics, alter the tone of a sound, increase loudness, and perform many other creative functions. Because of the wide range of uses for this tool, an understanding of compressor models and settings can be very useful. It often takes months or even years for audio engineering students to fully understand what compressors do and how to set them up correctly.

So what are compressors? The simplest definition is that they are automated volume faders. They work only by turning down the volume of what is run through them. Depending on the content of the audio signal that is being turned down and the "shape" of the volume reduction, different effects can be achieved through compression.

Before we get into the inner workings of compressors, it is important to understand the designs of hardware units. Knowledge of how vintage units work and which ones are best used in particular situations can speed up the mixing process and enhance the quality of the final product. Since the mid-2000s, a trend in audio plug-in development has been to emulate the expensive hardware that the big studios use. This emulation means that high-quality compressors are now available within the DAW. There are six main designs of compressors: optical, variable mu, diode bridge, voltage-controlled amplifier, pulse-width modulation, and field-effect transistor, each with its own sonic characteristics. We discuss them in the next section.

15.1 Hardware Designs

15.1.1 Optical

Optical compressors are the slowest and the smoothest sounding. They work by using a light and a light-sensitive device to dictate how much compression should be applied and at which speed. The brighter the light, the more compression there is. These compressors are very easy to use, as they have only one control: more or less compression. The attack and release speeds change, depending on the input volume. The more these compressors are driven, the faster are the attack and release. This characteristic allows for smooth compression because the compressors start slowly and ramp up to higher speeds as more compression is needed. Overall, these compressors have a "smooth" quality and are very efficient at gently leveling dynamics; thus they work particularly well for compressing vocals. The most famous optical compressors are the Teletronix LA2A (Figure 15.1) and the Tube Tech CL1B.

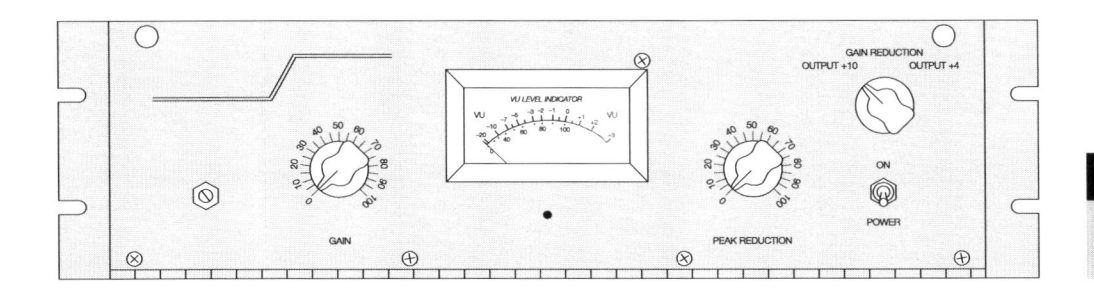

FIGURE 15.1

Teletronix LA2A leveling amplifier.

15.1.2 Variable Mu

Variable-gain compressors are little faster than optical ones, but still remain in the "lazy compressor" group. They have variable attack and release times, but no ratio control. Their ratio is increased as the compressor works harder. These types of compressors are mostly used for their sound, as they are colored pieces of hardware. They are also the most expensive type of hardware compressors available and are therefore rarer than other models. The most famous vari-mu compressors are the Fairchild 670 (Figure 15.2) and the Manley Vari-Mu.

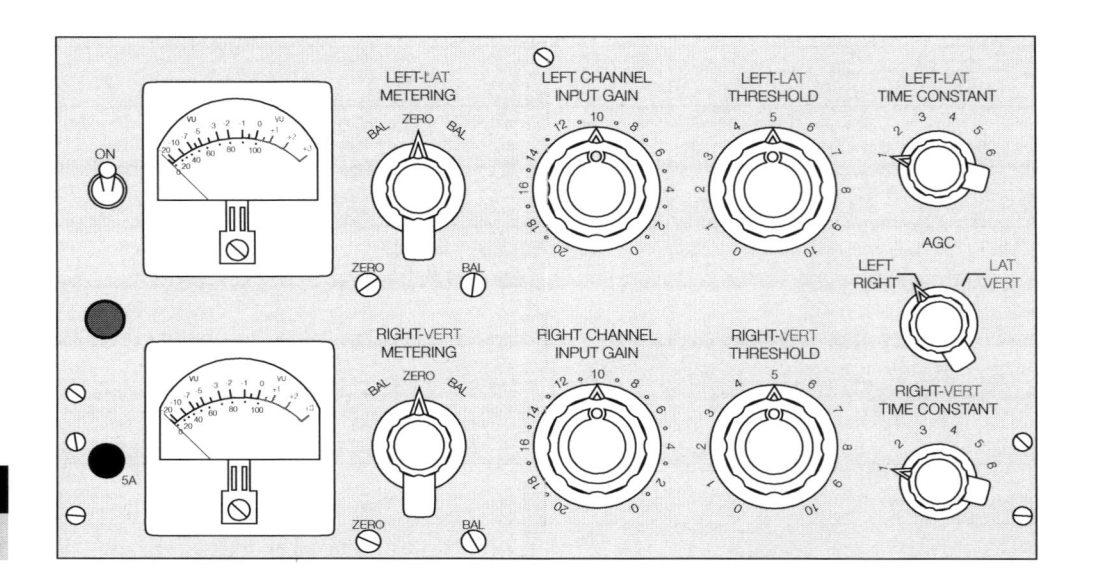

FIGURE 15.2 Fairchild 670 compressor.

15.1.3 Diode Bridge

This design is one of the rarest of hardware compressors. Diode-bridge units can be clean with low compression or colored when pushed hard. Gentle compression on these units can add snap to drums, but heavier settings will tend to soften transients as more distortion is introduced. They are very versatile in terms of attack and release speeds and can therefore be used for most applications. The most famous diode-bridge compressors are the Neve 2254 and the Neve 33609 (Figure 15.3).

FIGURE 15.3 Neve 33609 limiter/compressor.

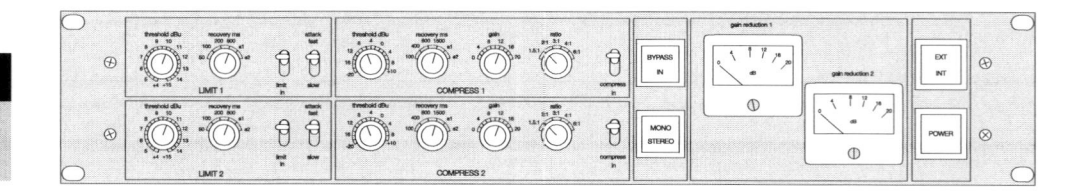

15.1.4 Voltage-Controlled Amplifier

Voltage-controlled amplifier (VCA) compressors are all-around clean units. They can compress fast and slowly, have different ratio settings, and are the most versatile type of compressors. Some units are more colored than others (the API 2500, for example), and some work faster than others (the DBX 160, for example). Overall, these compressors are meant to be used on just about anything; they perform well on single instruments and groups. Some famous units include the SSL G-Series Bus Compressor (Figure 15.4) and Empirical Labs Distressor.

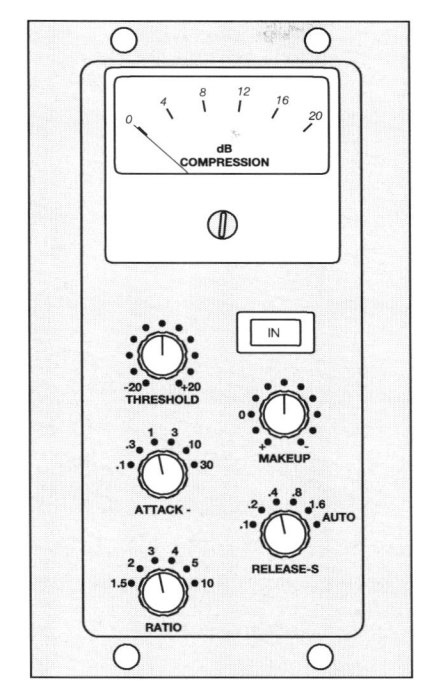

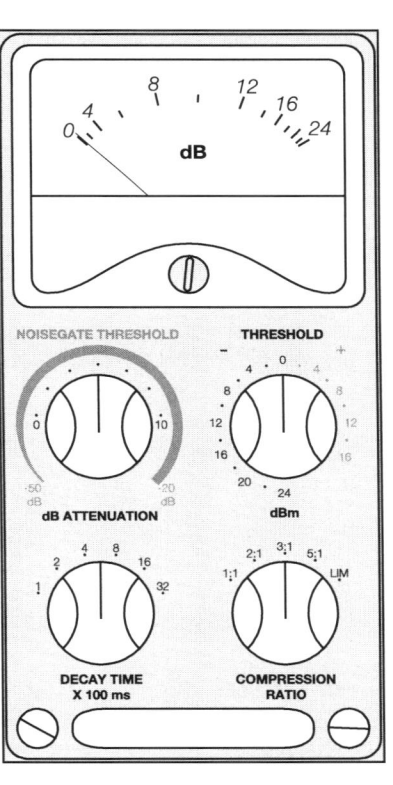

FIGURE 15.4

SSL G-series bus compressor.

15.1.5 Pulse-Width Modulation

Pulse-width modulation (PWM) compressors are similar to VCAs in the sense that they are versatile and work well on just about anything. They differ in their way of operation, as they are essentially a very fast gate (opening and closing in the MHz speed area). This fast turning on and off of the signal means that more or less sound gets through the compressor, ultimately reducing the output volume. The sound of PWM compressors is slightly smoother than that of VCAs. Famous units include the PYE compressors (for example, Figure 15.5) and Cranesong Trakker.

15.1.6 Field-Effect Transistor

Field-effect transistor (FET) compressors are aggressive. They are hardware limiters with the fastest attack possible, making them perfect when chopping peaks off is required. They can be very noticeable when pushed, but are great at gaining extra loudness. Even at slow settings, they still remain faster than other types of compressors. Famous units include the Urei 1176 (Figure 15.6) and API 525.

FIGURE 15.5

PYE 4060 compressor.

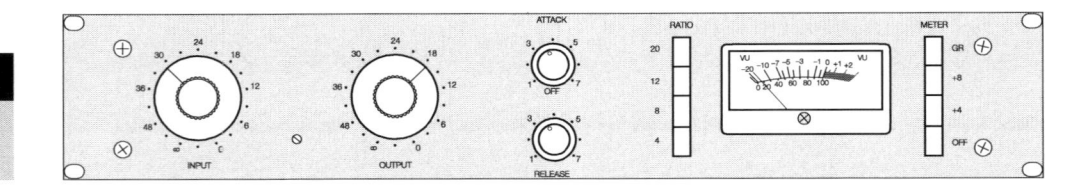

AUDIO PRODUCTION PRINCIPLES

FIGURE 15.6 Urei 1176 limiting amplifier.

15.1.7 Tape

Even though it is not technically a compressor, tape (Figure 15.7) is often considered as a great dynamics control device. Since driving a tape clips a signal and adds character, it is important to remember that this tool can also be used for dynamic control. The difference between tape and peak compression is that tape turns down only the high peaks via soft clipping and leaves the rest of the waveform untouched. This form of dynamics control means that the underlying instruments playing softly in the background do not get affected. Peak limiters (or compressors) work differently; they turn down the whole signal when their threshold is hit. The underlying elements get distorted and reduced in volume with each new peak triggering the compressor (adding attack and release times to limiters then complicates the subject matter even more!). Because of this phenomenon, mixing to tape differs greatly from mixing digitally. With tape, it is often desirable to mix in extra transients, since they will be "shaven off" when driven through the tape. When you are mixing digitally, the aim is often to get an already compressed and not overly dynamic mix. This step will ensure that the mix remains intact after mastering and will also translate well when compressed for radio.

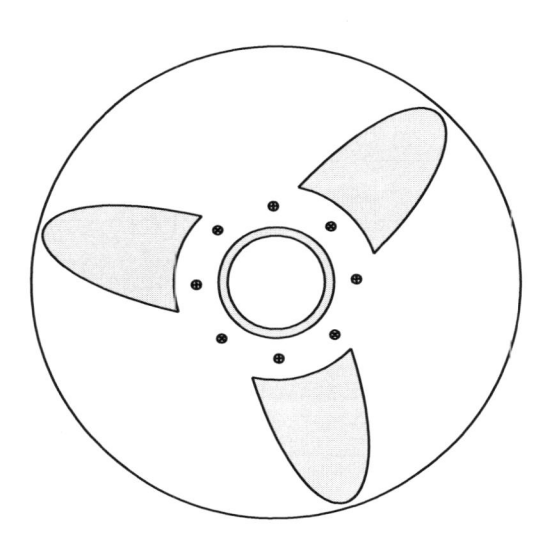

FIGURE 15.7 Tape roll.

The different controls on a tape machine (or more recently, on tape emulation plug-ins) can greatly affect the tone of a signal. The most common are the following:

15.1.7.1 Bias

Bias control can help with the fidelity of the tape machine (Audio 15.1 ▶). It is, however, possible to over- or under-bias the machine to achieve nonlinear effects such as a slight warming of the signal or distortion being added. Experimentation is essential in order to obtain the best results for the material you are working on.

15.1.7.2 *Tape Speed*

The tape speed used affects fidelity and frequency response (Audio 15.2 ▶). The three most common settings, measured in inches per second, are 7.5 ips, 15 ips, and 30 ips. The faster the speed, the higher the fidelity is and the flatter the frequency response. As speeds get lowered, the frequency response tilts toward a raised low end and slightly dulled high end. The genre of music and material being processed should dictate which speed is to be used. For example, a bass guitar could benefit from the stronger low end of 7.5 ips and slight reducing of string noise, while a full classical mix would require the cleaner tone of 30 ips speed.

15.1.7.3 *Tape Width*

The different tape sizes available have an effect on the quality of audio. Audio quality increases with tape width.

15.1.7.4 *Calibration Level/Fluxivity*

This setting relates to the type of tape being used (Audio 15.3 ▶). The settings commonly available are +0 dB, +3 dB, + 6 dB, and +9 dB, or can sometimes be displayed in nanoweber per meter (nW/m) values. While different tape types require different calibration levels, changing these types affect the saturation level: the higher the calibration, the more saturation.

15.2 Compressor Settings

15.2.1 Threshold

Threshold is the amplitude value of the input signal at which the compressor starts reducing gain (Figure 15.8). If the compressor is not already engaged and input signal is below the threshold, no compression is achieved. Once the input signal has reached the threshold, the compressor starts reducing gain. At this point, even if the input signal falls back under the threshold, the compressor will keep reducing gain until the compressor is fully released.

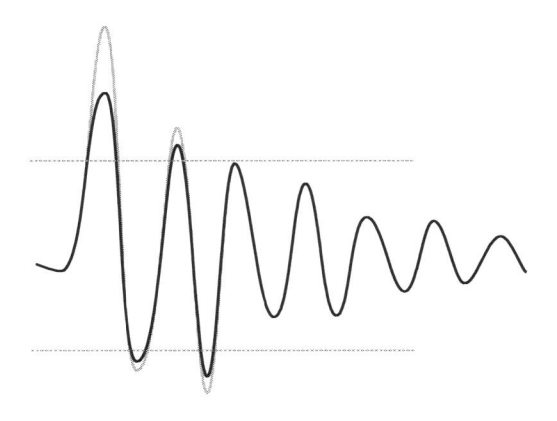

FIGURE 15.8
Compressor threshold.

15.2.2 Ratio

The ratio sets the input to output ratio when the compressor is fully engaged (Figure 15.9; Audio 15.4). A ratio of 2:1 means that for every 2 dB that go in, only 1 dB will come out. Ratio directly relates to the perceived "size" of the signal and "aggressiveness" of the compressor. High ratios can reduce dynamic range very effectively. It can be useful to separate ratios into the following four groups:

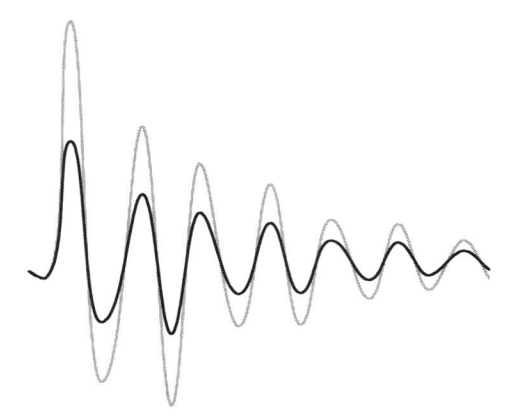

- 1:1 to 2:1 for subtle dynamic control and groups of instruments processing (glue)
- 2:1 to 4:1 for medium dynamic control and individual instrument processing (tone)
- 4:1 to ∞:1 for heavy dynamic control (limiting) and individual instrument processing (tone)
- Negative ratios for special effects

FIGURE 15.9 Compressor ratio.

15.2.3 Attack

Attack is a time constant that dictates the speed at which the compressor reduces the gain of the signal going in (Audio 15.5). The exact definition of this control largely depends on how the compressor is designed. Attack could be the time it takes to reach two-thirds of the ratio, the time it takes to reach 10 dB of gain reduction, or anything else that the manufacturer has designed it to do. An important point is that because the attack is a time constant (rather than a period of time), the actual speed of the compression is also dependent on other factors such as ratio, knee, and input-sound transient gradient. For example, if the attack is the time it takes the compressor to reach two-thirds of the ratio, the higher the ratio the faster the compression speed, since the unit will compress more in an equivalent amount of time.

As a general rule, fast attack settings will cut the initial transient, which sometimes sounds fatter, choked, or distorted (the tone achieved depends on the other controls). Note that aliasing can sometimes be introduced with very fast attack settings (but this issue can be reduced if the compressor works in oversampling mode). Slow attack settings will let the initial transient pass through intact, and thus can add punch but can also sound thin at times. Note that longer attack times reduce the effect of the release time. Once again, the compressor settings all interact with one another, hence the range of tones that can be achieved with this control.

15.2.4 Release

Release dictates the speed at which the compressor will stop reducing gain (Audio 15.6 ▶). Release is also a time constant and its value also depends on the design of the compressor. For example, release could be the time it takes to recover the signal by 10 dB, or the time it takes to go from full ratio back to 1:1 (no compression). Understanding that release is not a period of time helps in understanding why the actual speed of the release also depends on the ratio, release type, and current input volume.

The release time usually gives more predictable results in tone. Fast release gives more "attitude" or "aggressivity" to the sound, and yield higher volume level. You should be careful with very fast release times, as pumping and distortion can be introduced. This result is even more apparent in low-end instruments. Slow release times can sound more natural and keep the original tone, but too slow a release and the instrument may sound a little "lazy" or "soft." As a general rule, releasing in time with the pace of the song will give the most transparent results.

15.2.5 Hold

Some compressors feature a hold function, which works in conjunction with the release time to achieve different release curves. The hold value is often the time it takes the release to go from its slowest setting to the set value. The release therefore gradually reaches the value set over the period defined by the hold value.

15.2.6 Knee

The knee is the shape of the gain reduction or, put simply, the attack curve. It is the curve that the input signal will follow until it reaches full compression ratio (Audio 15.7 ▶). Hard knee (Figure 15.10) will give hard and punchy results, while soft knee (Figure 15.11) is used for softer and smoother applications where the compressor needs to be more transparent.

FIGURE 15.10
Hard knee.

15.2.7 Linear/Logarithmic Mode

Some compressors offer the option of working in either of these modes for the attack, release, or both (Audio 15.8 ▶). This option directly affects the smoothness of compression. Linear mode often gives a more aggressive and unnatural compression sound, while logarithmic mode is smoother and less intrusive.

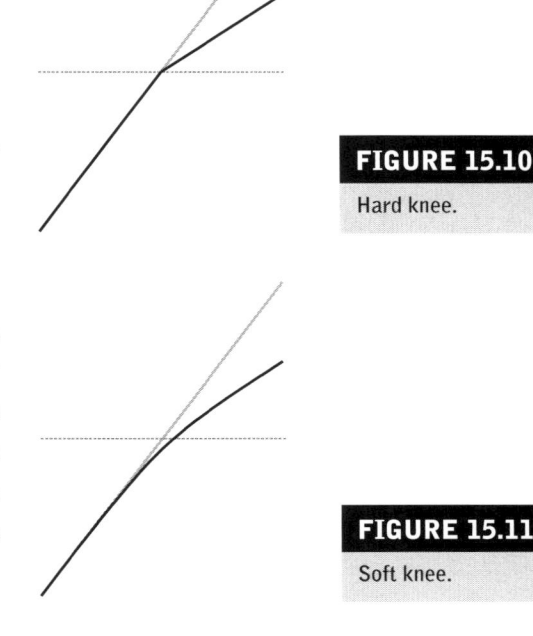

FIGURE 15.11
Soft knee.

15.2.8 Over-Easy, Auto-Release, and Auto-Fast

Some compressors offer automatic attack and/or release, usually achieved by detecting the envelope of the signal and adjusting the attack and release settings to match. This process ensures that the compressor is fast enough to catch all transients, and releases following the envelope of the sound. While these settings are useful for general dynamic control duties, they do not work well when a particular effect needs to be achieved (punch, fatness, etc.).

15.2.9 Makeup Gain

Since compressors reduce the volume of audio being run through them, an output gain function is needed in order to match in/out levels. This step is very important for an engineer being able to judge whether or not the compression settings are appropriate. Just as with any other signal processor, being able to bypass the compressor and stay at an equal volume level is critical.

15.2.10 Side-Chain Input

It is possible to use an audio signal to trigger the compressor but have its effect applied to a different audio signal. In this case, the signal to be used for triggering and dictating how the compressor behaves should be routed to the side-chain input (Figure 15.12). This technique is often used in supermarkets to enable the voice of workers talking through the PA to reduce the volume of the background music.

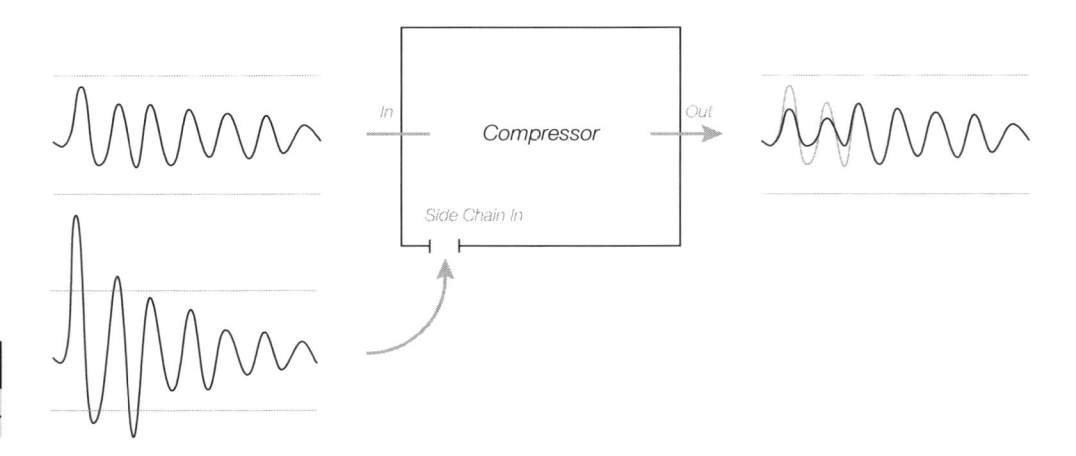

FIGURE 15.12
Compressor side-chain input.

15.2.11 Stereo Link

Some hardware and plug-in compressors can be used as either dual-mono (Figure 15.13) or stereo units (Figure 15.14). Although they often have only one set of controls, dual-mono units work as two separate compressors, one for each channel of audio. Stereo units will still run as two compressors (one for each channel), but the detection circuit for each side takes its input from both channels of audio. Thus a loud transient on the left channel will trigger compression on both the left and right channels, and vice versa (Audio 15.9 ▶).

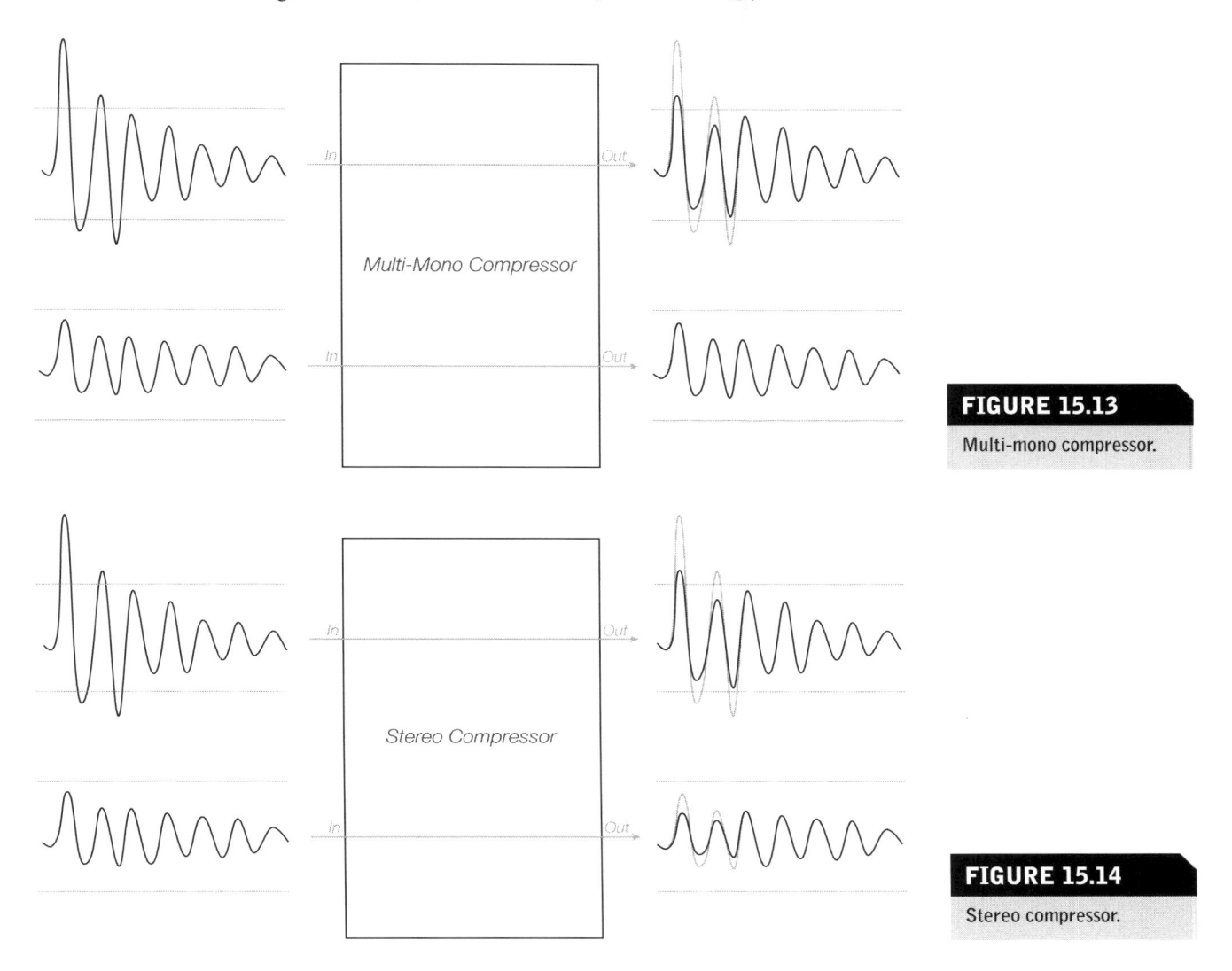

FIGURE 15.13
Multi-mono compressor.

FIGURE 15.14
Stereo compressor.

There are advantages to linking and unlinking compressors, and the material to be compressed should dictate the most appropriate setting. For example, dual-mono compressors should not be used on tracks that have been recorded by using coincident pairs of microphones such as an XY or Blumlein technique. Because of the nature of such recording (using mostly amplitude differences to create the stereo image), reducing the volume differences between the two channels will result in a more mono output from the compressor. On tracks that have been recorded by using a spaced pair of microphones, though, this method can be useful in creating a more solid stereo image by evening out the dynamics on each side, while keeping the time differences intact.

Some compressors also provide the feature of transient and release linking separately. This feature allows the compressor to detect transients separately (unlinked), therefore making the compression attack smoother when the shifts in level are too fast. Similarly, the release can be unlinked to achieve maximum loudness when consistency in the stereo spread is not crucial.

15.2.12 Lateral/Vertical (Mid-Side)

Some compressors can operate in mid-side mode rather than left-right. While it was originally used in the Fairchild 670 for vinyl disc mastering, this method can create interesting effects on some sound sources.

15.2.13 Detection Type

Different compressors vary in the way they detect the incoming audio signal (Audio 15.10 ▶). Some compressors detect individual peaks, a capability that makes them react faster and more aggressively, but could lead to undesirable pumping. Others work by detecting the average level (RMS detection) of the audio coming in and thus are often more natural-sounding units. It is commonly accepted that peak detection works well on single instruments (especially drums and other instruments with sharp transients), while RMS detection gives better results for groups and mix-bus compression.

15.2.14 Detection Circuit

Modern compressors use either feed-forward (Figure 15.15) or feedback (Figure 15.16) detection circuits (Audio 15.11 ▶). Feed-forward compressors react to the original signal being input into the unit, while feedback designs place the detection circuit after the signal has been compressed. In use, feed-forward compressors tend to sound more "modern" and "immediate," while feedback units are "smoother" and more "musical," as they are constantly adjusting to the signal being compressed.

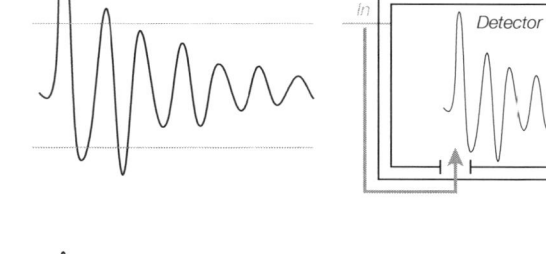

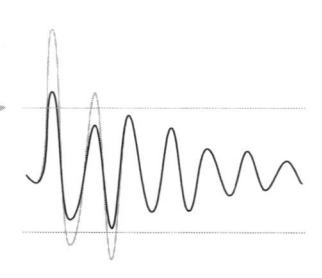

FIGURE 15.15
Feed-forward compressor design.

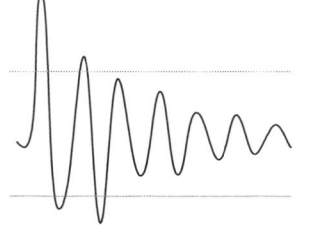

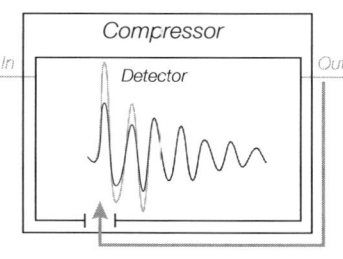

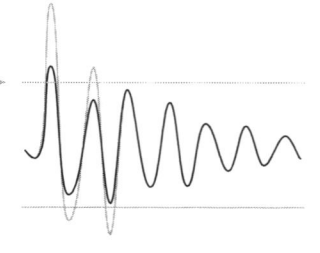

FIGURE 15.16
Feedback compressor design.

15.3 Why Use Compressors?

Depending on the settings used, compressors can give very different results. It is generally accepted that compressors can work on micro-dynamics (altering the inner dynamics or peaks of individual instruments) and macro-dynamics (altering the overall loudness of longer passages). On individual instruments, compressors can do one of two things: decrease dynamic range by cutting the peaks of the signal down (compressing for fatness), or increase dynamic range by letting peaks through and reducing the low-level signal further (compressing for punch). In this case, the compressor is used as a tone effect, which will largely depend on the attack setting of the compressor.

A fast attack and release (Figure 15.17; Audio 15.12 ▶) will cut any transient going through and therefore reduce the dynamic range. By doing so, the compressor is decreasing the gap between low-level signal and loud parts, resulting in a fatter sound.

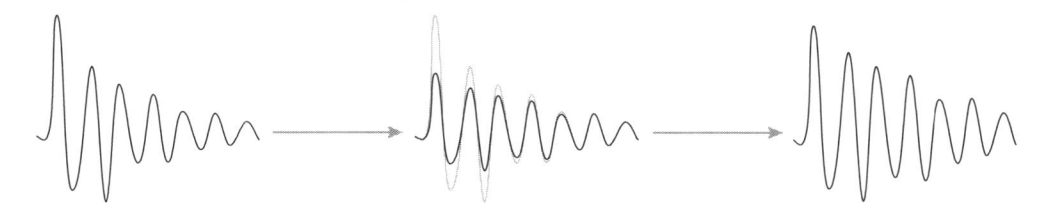

FIGURE 15.17
Fast-attack, fast-release settings.

A fast attack and slow release (Figure 15.18; Audio 15.13 ▶) will even out the dynamics but retain the overall original dynamic curve of the signal. Since the whole signal will be compressed, both the transient and sustain parts will be reduced in volume, but at different rates.

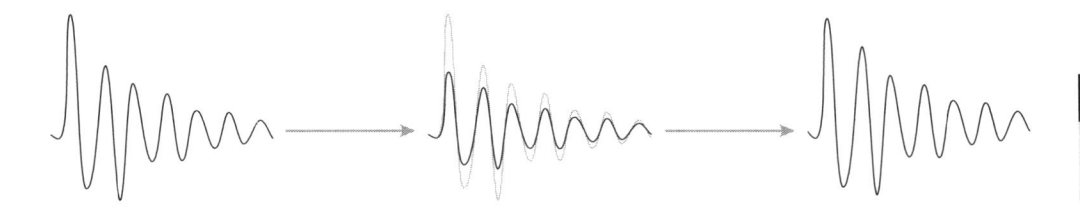

FIGURE 15.18
Fast-attack, slow-release settings.

A slow attack (Figure 15.19; Audio 15.14 ▶) will let the transients through, then reduce the volume, thus increasing the gap between the low-level signal and the loud parts. This process effectively increases dynamic range, resulting in a "punchier" sound.

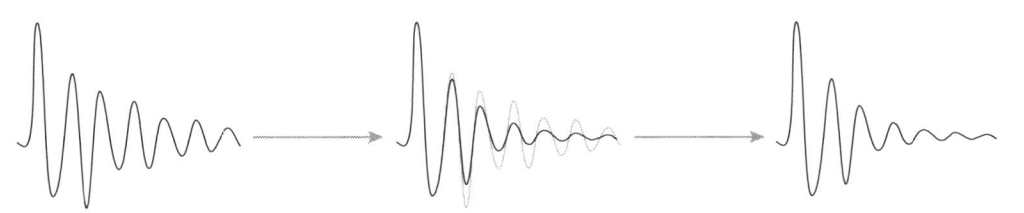

FIGURE 15.19
Slow-attack setting.

For an alternative way of achieving fatness or punch, see Box 15.1.

> **BOX 15.1**
>
> *Transient designers are special processors based on envelope followers that allow for very simple control of the attack and sustain of instruments. These processors have controls for boosting or reducing attack and release of a sound. They can be great tools for achieving fast results when setting up a compressor may break an otherwise fast-paced workflow. Boosting the attack or reducing the sustain on these processors sounds similar to a compressor set with slow attack. Reducing the attack or boosting the sustain on these processors sounds similar to a compressor set with fast attack. Depending on how much processing is applied, different effects can be achieved (Audio B15.1 ▶). For example, reverb can be raised or lowered by using the sustain control, and overall RMS level can be raised by slightly lowering the attack control.*

On groups of instruments, compressors can be used somewhat differently. They can give the illusion of glue and polish to the group of instruments, as every sound will trigger the compressor which acts on the whole group. Where each instrument affects the dynamics of others, a sense of togetherness is achieved through this interaction. Note that while wide-band compressors will glue instruments by triggering overall level changes in the instruments they process, it is also possible to use multiband compressors to achieve glue on a frequency by frequency basis. For example, where all the different tracks of a multi-tracked drum kit would trigger a wide-band glue compressor inserted on the drums bus, inserting a multiband glue compressor on a mix bus would glue individual frequencies together.

Compressing for loudness has also been a trend since the 1990s. Achieving loud masters is often vital in order to compete on the commercial market. If loudness is sought after, compressors must be used at all stages of the mix with loudness (i.e., reduced dynamic range) in mind. Loudness cannot be achieved by simply adding a limiter on the master; it must be planned ahead. For example, you might compress a snare drum, then the drums group, then the drums and bass group, and finally the master. This process will ensure that no compressor is working too hard but that high compression is achieved overall.

For increasing the perceived loudness without altering volume or using compression, see Box 15.2.

> **BOX 15.2**
>
> *Because multiple bursts of sounds occurring in a short amount of time are perceived to be louder than the same sound played continuously (at the same loudness), it is possible to make a sound come out in the mix without raising its volume or using compressors. Experimenting with stutter FX can help make an instrument more noticeable in the mix. Lengths of sound bursts between 50 ms and 200 ms tend to work best for this effect.*

15.4 Setting Up a Compressor

1. Start by setting all of the controls to the following: highest ratio, fastest attack, fastest release, hard knee.
2. Bring the threshold down until a healthy amount of gain reduction is achieved (around 20 dB is usually enough). Make sure not to compress too much, as this technique does not always succeed when the compressor is working too hard.
3. Slow the attack down until the desired amount of transient is let through uncompressed. Having the compressor work really hard when you are setting the attack means that a slower attack can be heard instantly.
4. Slow the release down until the following occurs:
   - Any distortion caused by the heavy compression disappears.
   - Any pumping artifacts disappear.
   - The compressor releases in time with the pace of the material being compressed, but not any more slowly, as the next transients hitting the compressor will be choked. The key is to have the compressor fully released before the next transient event.
5. Set the ratio in accord with the type of compression to be achieved: high ratios for heavy compression, low ratios for subtle compression. It is difficult to achieve transparent compression by using ratios above 4:1.
6. Set the knee in accord with the type of compression to be achieved: hard knee for aggressive compression, soft knee for smooth compression.

7. Adjust the threshold to achieve the desired amount of gain reduction. Having set every control with the compressor working quite heavily, you can keep the sonic characteristics of the compressor but they will be less intrusive when you are reducing the amount of gain reduction.
8. Adjust the attack and release settings to ensure that the compressor still acts as desired. While this effect is often harder to hear, the final settings for the attack and release should not be too far off from what has already been set.

This technique is derived from the well-known method of attack, release, ratio, threshold (ARRT); the difference is that the starting point is full compression rather than no compression (Audio 15.15 ▶; Box 15.3).

> **BOX 15.3**
>
> *Compressors take time to understand and set up. Using presets in plug-in compressors is often a good starting point for understanding how the different controls can achieve different sounds. The "kick punch" preset on your DAWs compressor will almost always add punch to kicks, and the "smooth vocal" preset will often reduce the dynamic range of the voice. When using presets, you will always need to adjust the threshold and makeup gain. There is no way for whoever created the preset to know how loud your particular kick or voice is. Therefore, while such elements as the attack, release, ratio, and knee might be set properly for the desired effect, the threshold and gain will always be wrong.*

15.5 Parallel Compression

Sometimes called New York compression, this way of working with compressors is most useful when fatness is to be achieved without sacrificing the original transients of the signal being processed (Audio 15.16 ▶). This method is set up by duplicating the original signal and highly compressing one of the duplicates (Figure 15.20). High-ratio, fast-attack, and fast-release settings should be used to ensure that none of the transients are let through. Blending this highly-compressed duplicate with the uncompressed original effectively keeps the best of both tracks: the transient of the uncompressed signal is heard first, then the sustain of the compressed signal. This technique is often used on drums, but can also work quite well on very dynamic vocals, or even full mixes. Interesting results can also be achieved by using brick-wall limiters or hard clippers in parallel instead of compressors.

FIGURE 15.20
Parallel compression.

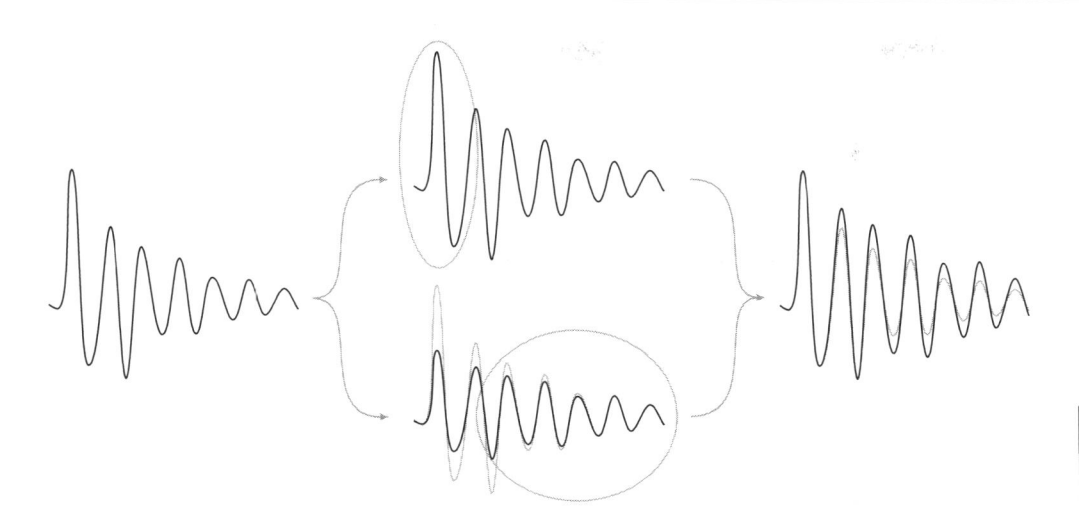

This technique can then be pushed further by applying extra processing on the duplicates. Once you have achieved an appropriate blend, the loud and dynamic part of the signal is given by the original uncompressed sound, and the low-level sustain part is given by the parallel compressed signal. This blend means that the overall signal heard will be constantly shifting from the original to the duplicate. At this point, applying different EQ, saturation, and other effects to each will add extra movement to the sound. For example, the original could be brighter, while the parallel duplicate could be more saturated and dark. In this case, the sound will be constantly shifting from bright transient to thick sustain.

15.6 Serial Compression

It is possible to use two or more compressors in series to achieve high amounts of gain reduction without hearing too much compression artifacts (Figure 15.21; Audio 15.17). The rule is to start with the more aggressive compressors, and follow with the gentler ones. The first compressor used should be set as a limiter: high ratio, fast attack, and fast release. FET compressors such as the Urei 1176 work perfectly for this task. The second compressor should have a lower ratio, slower attack, and slower release. Optical compressors such as the Teletronix LA2A work well for this task, as they are very gentle. In most cases, two compressors should be enough to achieve high gain reduction without making the artifacts associated with heavy compression audible. The theory behind this technique is that using two compressors with a gain reduction of 5–7 dB each will sound a lot more transparent than using one single compressor achieving 10–14 dB of gain reduction. To ensure that fewest artifacts are heard, adding gain automation before the compressors as "manual" compression may be necessary. This step will allow the compressors to work even less, as some of the dynamics will have already been taken care of with automation. This technique is often used on pop vocals, but can also work on other very dynamic instruments.

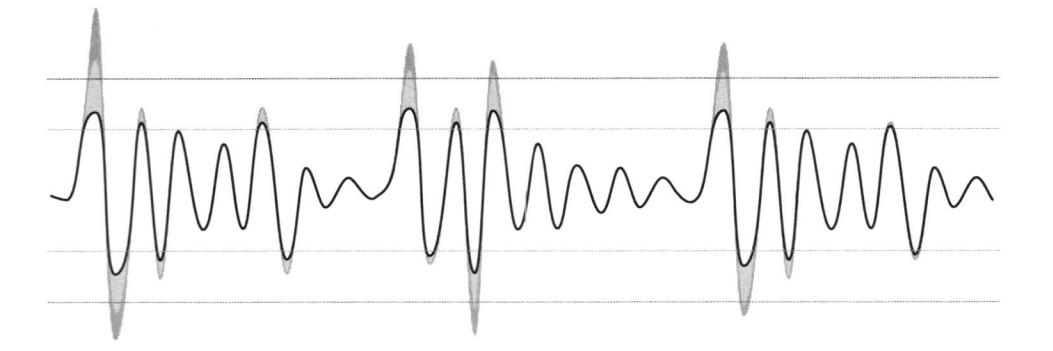

FIGURE 15.21
Serial compression.

15.7 Equalizer-Compressor-Equalizer

Depending on their purpose, EQs can be inserted before or after compression. If the EQ is "fixing" the sound source (removing low-end rumble or a resonant frequency, for example), it should be inserted before compression, to ensure that the compressor is reacting only to signal that is meant to be there rather than having it "pump" for no apparent reason. Compressors can sometimes change the tone of a signal by making it duller. It is sometimes necessary to insert an EQ after this process to add brightness back and keep the original tonal aspect of the sound (Figure 15.22; Audio 15.18 ▶).

FIGURE 15.22
Processing order: EQ, compression, EQ.

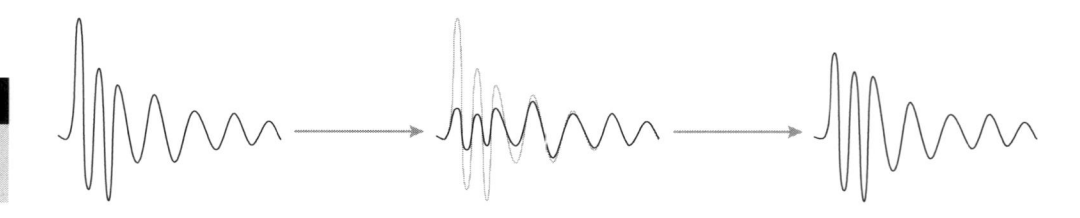

15.8 Increasing Reverb

Compressing an ambient microphone such as the overheads of a drum kit with a fast attack setting can increase the perceived reverb level. Since the tail of each sound will ultimately be raised by the compression, the volume of the room ambience will also be raised.

15.9 Side-Chain Input Uses

It is possible to use the side-chain input of a compressor to separate the signal being detected from the signal being affected. This tool can be powerful for helping separate instruments in the mix. Common examples include the following:

- PA systems where voice must be heard over the music. Sending the voice to the side-chain input of a compressor inserted on the music channel means that every time the voice is active, the music is compressed (Figure 15.23; Audio 15.19).

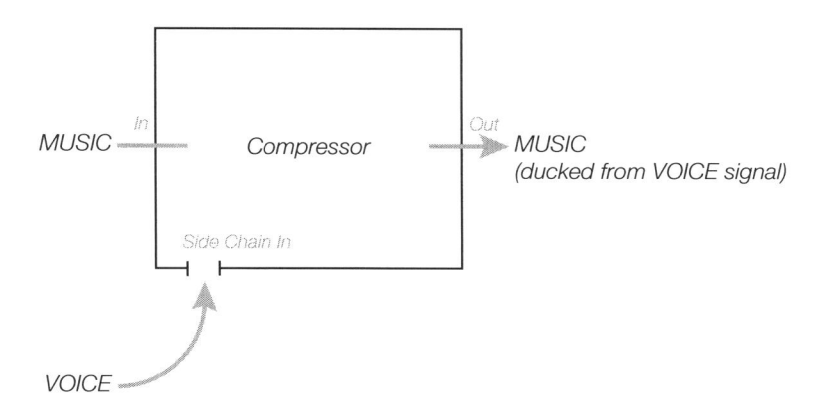

FIGURE 15.23
Ducking music from voice signal with a compressor.

- Kick/bass separation. Because kick and bass often "live" in the same frequency area, this technique is very useful for separating them. Sending the kick to the side-chain input of a compressor inserted on the bass means that every time the kick hits, the bass is compressed (Figure 15.24; Audio 15.20). This technique can also be achieved by using a multiband compressor and reducing only the fundamental frequency of the kick on the bass track. This way the bass retains all of its power through the frequency spectrum except where the kick "lives."

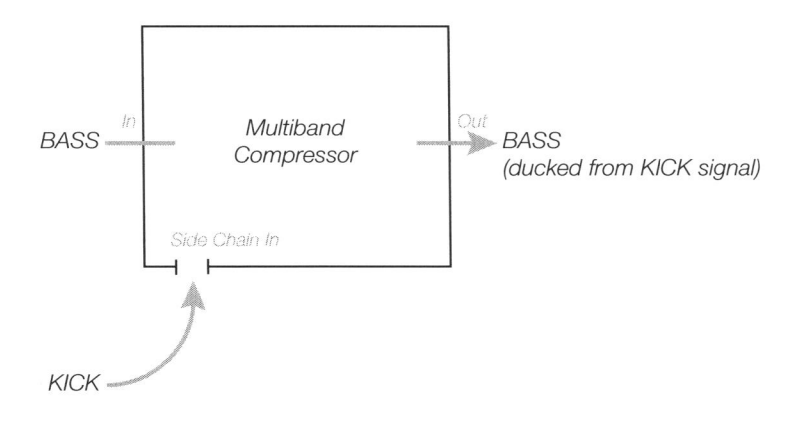

FIGURE 15.24
Ducking bass from kick signal with a compressor.

- Voice/guitar separation. A similar technique can also be used to separate the voice and guitars. It is often preferable to send the voice to the side-chain input of a multiband compressor inserted on the guitar track, with the mids band only compressing (Figure 15.25; Audio 15.21 ▶). This approach means that instead of turning the whole guitar track down when the voice is active, only the mids (where the voice "lives") are being compressed. This technique works even better if the multiband compressor is operating in mid-side mode and compressing the center channel only, leaving the sides untouched. A similar effect can be achieved by using a gate inserted on the voice, and triggered by the guitar. In this case, the voice is always reduced in volume unless the guitar is playing, in which case the volume of the voice is boosted

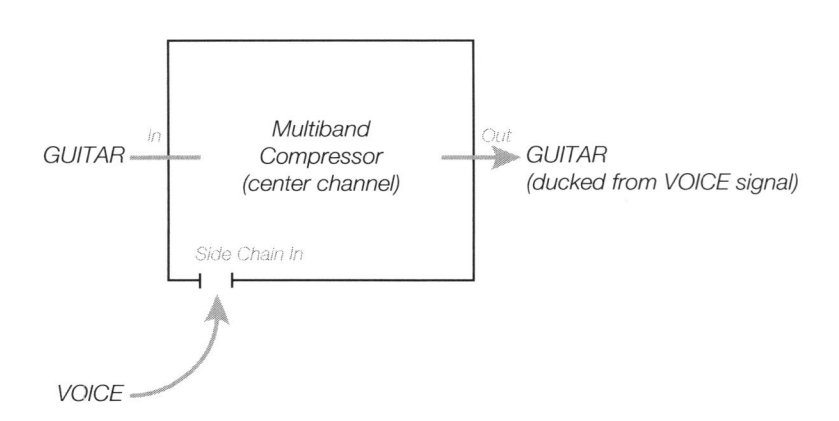

FIGURE 15.25
Ducking guitar from voice signal with a compressor.

- Tightening kick/bass; tightening lead/backing vocals. While this technique does not work as well as manual editing, it is possible to tighten up the performance of kick and bass by inserting a gate on the bass and have it triggered by the kick. This method will effectively let only the bass play when the kick is present. While attack should be set as fast as possible, tweaking the hold and release times will help keep the original length of the bass notes. The same technique can be applied to lead and backing vocals, where the gate is inserted on the backings, but triggered by the lead. In this case, all the controls should be kept fairly fast so the ends of words on the lead track close the gate on the backings (Figure 15.26; Audio 15.22 ▶).

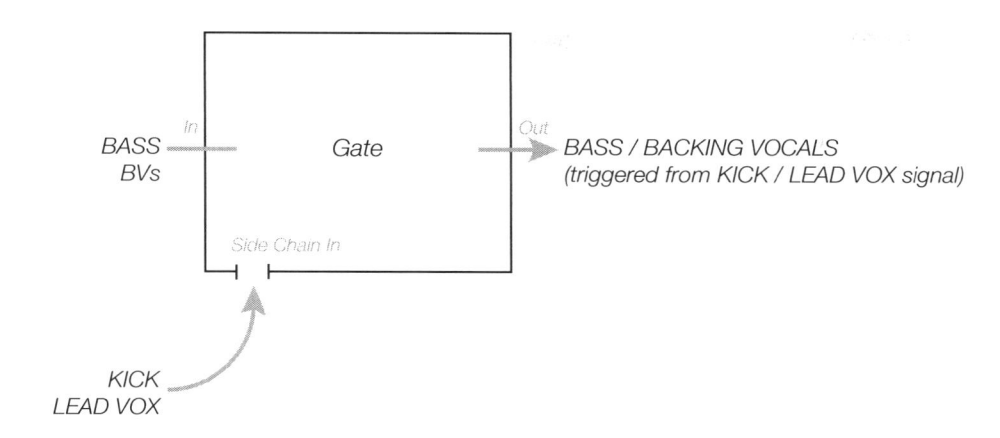

Compression, Expansion, and Gates

FIGURE 15.26
Triggering bass or backing vocals from kick or lead voice signals with a gate.

- Creating large-sounding snare drum. This technique uses a gate or expander on the room microphones of a live drum kit, triggered from the snare (Figure 15.27; Audio 15.23 ▶). In this scenario, each time the snare hits, the room microphones are getting louder, giving the impression of a larger and more reverberant snare.

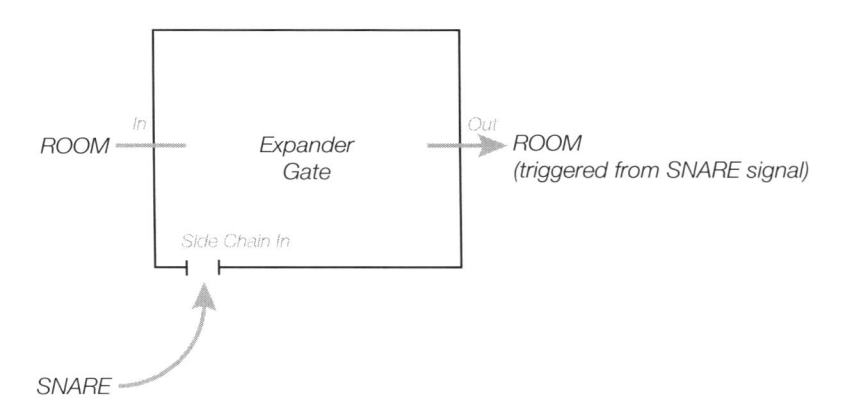

FIGURE 15.27
Triggering room microphones from snare signal with an expander or gate.

- Achieving extra glue on the mix-bus compressor in music that contains a lot of low-end information. Using a simple HPF or reducing the volume of the low end by using a low shelf in the side chain will force the compressor not to react to bass frequencies, and therefore glue the rest of the frequency spectrum better (Audio 15.24 ▶). This technique can be pushed further by boosting high mid-frequencies to reduce overall harshness by using the same compressor, for example. Working

this way somewhat emulates what a multiband compressor would do by reacting to particular frequencies more than others.
- Keeping the original transient of a kick drum being compressed with fast attack. If you use an LPF on the side chain of such compressor, the kick drum will still let the original transient pass through, but will react a lot more quickly once the low-end resonance of the kick has started, therefore giving a fat but punchy kick (Audio 15.25 ▶).
- Snare/hi-hat separation. In order to keep the snare hits as clean and punchy as possible, insert a compressor on the hi-hats triggered by the snare to duck the hats each time the snare is hitting (Figure 15.28; Audio 15.26 ▶). This effect does not work with all rhythms, so make sure that the hats are not unnaturally pumping. In the event that the hats and snare are too far out of time, this method can be great for masking the timing issues. A variant of this technique involves compressing the hi-hats by using the kick as the trigger. Another variant involves using an LPF (with side-chain input linked to the filter cutoff) instead of a compressor, a step that will create movement in the hi-hat frequency rather than lower its volume.

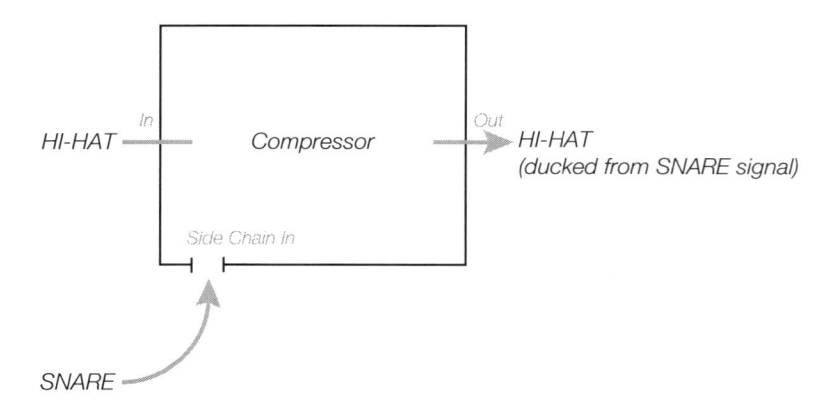

FIGURE 15.28

Ducking hi-hats from snare signal with a compressor.

- Effectively gating tom tracks containing a lot of bleed. In such a scenario, the toms should be fed into the side-chain input of their own gate, but equalized so that their resonant frequency is highly boosted (Figure 15.29; Audio 15.27 ▶). This step will fool the gate into thinking that the toms are a lot louder than the bleed.

Compression, Expansion, and Gates

FIGURE 15.29
Triggering toms from an equalized version of themselves with a gate.

- Effectively gating a snare track containing a lot of bleed. In this case, placing the gate on the snare-top track but triggering it with the snare-bottom microphone can work very well (Figure 15.30; Audio 15.28 ▶). High-pass-filtering the side-chain input to ensure that no kick bleeds onto the snare bottom will also help.

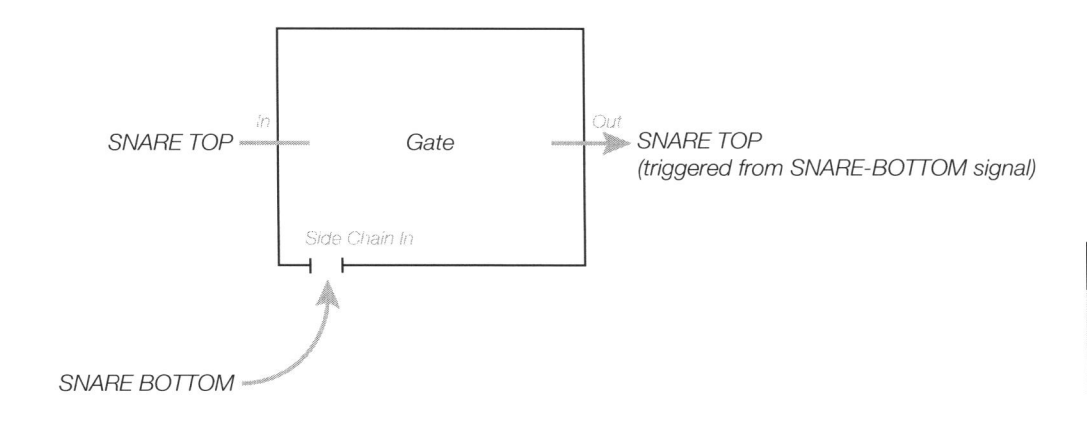

FIGURE 15.30
Triggering snare-top microphone from snare-bottom with a gate.

- Creative pumping effects in dance music. The kick drum is often fed into the side-chain input of a compressor placed on various instruments in dance music to create the very recognizable four-to-the-floor pumping effect (Figure 15.31; Audio 15.39 ▶). Using MIDI notes instead of the kick to trigger the compressor will allow for much more control over the pumping. In this case, the MIDI notes can be placed slightly before the kick to ensure that the compressor is already reducing the volume and give a sense of urgency to the pumping. The MIDI notes can be quantized and swung to give more feel to the pumping. The MIDI notes can even vary in velocity to add life to the pumping (if velocity is linked to the volume of the virtual instrument, of course).

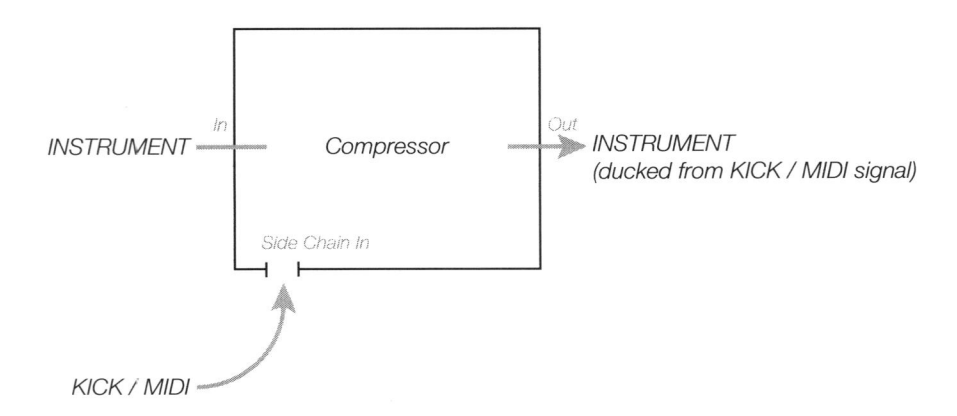

FIGURE 15.31
Dance-music pumping-effect example.

- Snare/Synth/Guitar interactions. Using an LPF on the guitars or synth track (with side-chain input linked to the filter cutoff) will allow the snare to trigger the closing of the filter and therefore punch through the mix better (Figure 15.32; Audio 15.30 ▶). This effect can be pushed quite hard for more creative applications.

FIGURE 15.32
Triggered filtering of top end out of synth or guitar from snare with a multiband compressor.

- Voice/reverb/delay separation. As mentioned in Chapter 10, "Vocals," inserting a compressor on the reverb or delay return and triggering it from the dry vocals can help clean up the vocals and make them more intelligible (Figure 15.33; Audio 15.31 ▶). This effect works even better when the trigger signal (the dry voice) is compressed before being sent to the side-chain input of the compressor placed on the reverb/delay return.

FIGURE 15.33
Ducking reverb return from compressed dry signal with a compressor.

- Guitar/hi-hats groove. To ensure that these two instruments are grooving together, insert a compressor on the guitars, and feed the hi-hats to its side-chain input (Figure 15.34; Audio 15.32 ▶). Light compression will make the guitar move in time with the hi-hats.

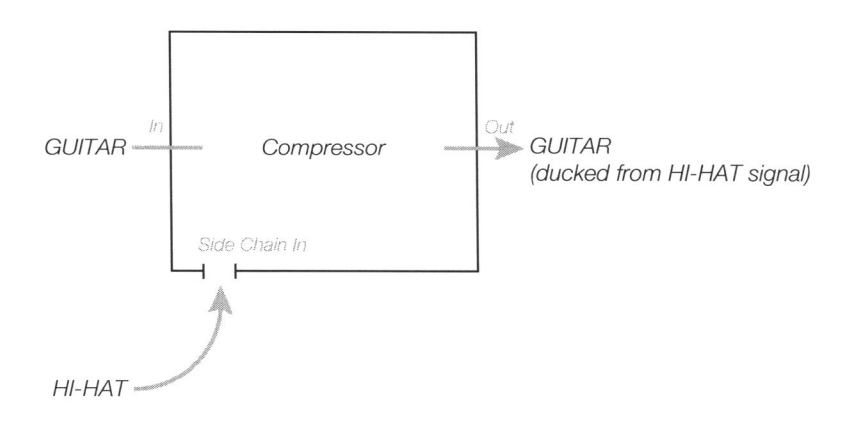

FIGURE 15.34
Ducking guitar from hi-hat signal with a compressor.

- Kick/snare separation. Using a multiband compressor inserted on the kick and triggering the top-end compression by the snare will allow for the definition of the snare to be fully present while keeping the weight of the kick for support (Figure 15.35; Audio 15.33 ▶).

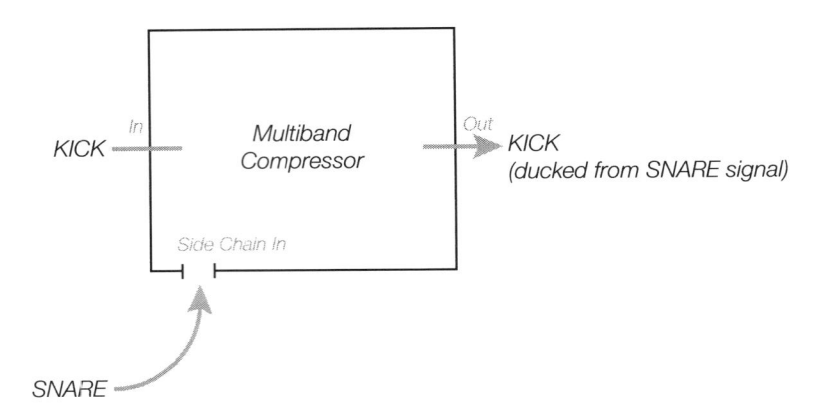

FIGURE 15.35
Ducking kick from snare signal with a compressor.

- Cleaning up the low end of reverbs. Using an HPF on the reverb return (with side-chain input linked to the filter cutoff) will allow the kick and bass to trigger the closing of the filter, therefore removing low frequencies when they are present (Figure 15.36; Audio 15.34 ▶). This method will permit the full frequency range to be present on the reverb return when no other low-end instrument is present.

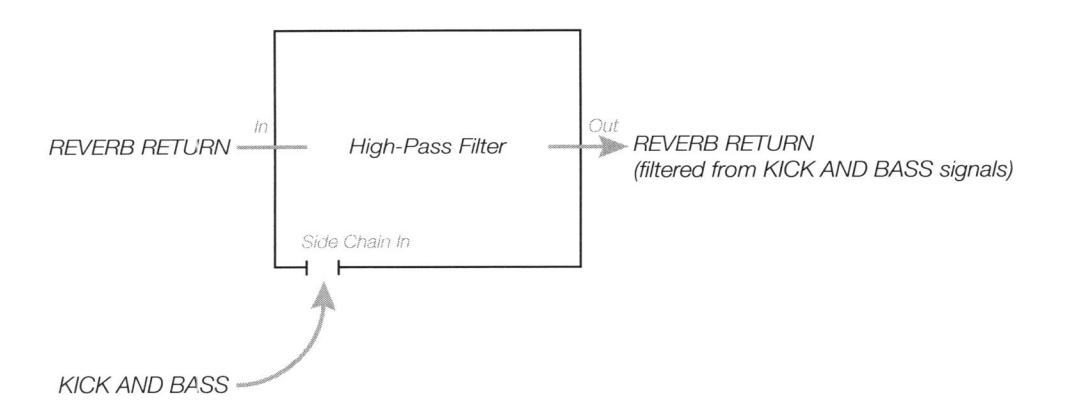

FIGURE 15.36
Ducking reverb return from kick and bass signals with a multiband compressor.

- Brightening a snare/emulating-snare bottom microphone. The aim in this case is to use the snare as a trigger for white noise. You must first create a blank track with printed white noise. After inserting a gate on that track, use the original snare to trigger the gate open (via its side-chain input) and therefore adding noise to each snare hit (Figure 15.37; Audio 15.35 ▶).

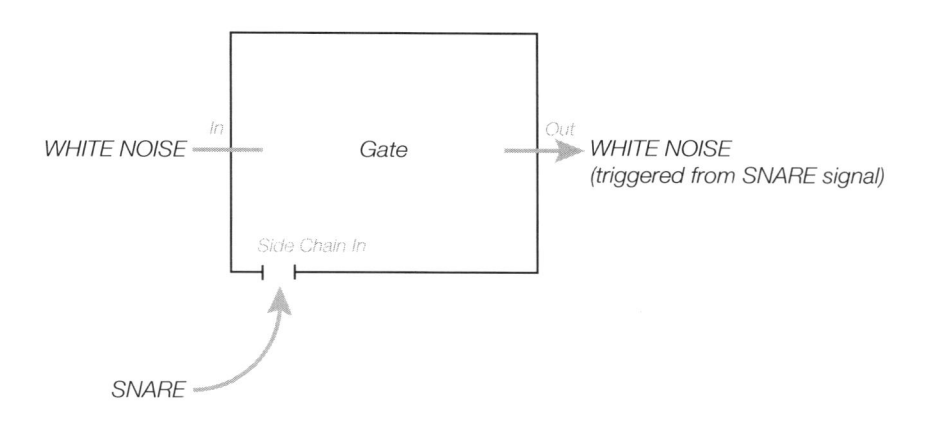

FIGURE 15.37
Triggering white noise from snare with a gate.

- Automatic ducking of artists when not recording. This technique is most useful when recording guitar players, but can be applied to any artist who plays all the time and disrupts the communication lines in the studio. First, a blank track needs to be created and filled with a steady tone for the duration of the song. This track should then be fed into the side-chain input of an expander placed on the artist's track. The threshold needs to be set at a point where the expander is closed when the DAW is stopped and opened when in playback mode. Since the printed tone triggering the expander will play back only when the DAW is playing/recording, the expander will automatically let the full volume on the artist's track when recording, but lower it when the DAW is stopped (Figure 15.38).

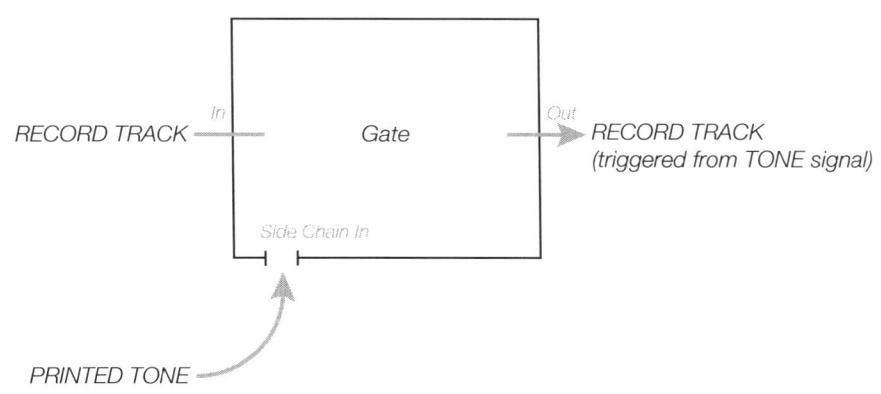

FIGURE 15.38
Ducking recording input signal from printed tone with a gate.

15.10 De-Essers

There are two kinds of de-essers: wide band and split band. Wide-band de-essers are regular compressors with an EQ boost applied to the signal being detected by the compressor (Figure 15.39). Thus the compressor reacts more to a particular frequency. In the case of vocals, the frequency boost is set to around 7 kHz to ensure that sibilance triggers the compressor.

Split-band de-essers work slightly differently, as the frequency selected by the user is the only frequency being reduced (Figure 15.40).

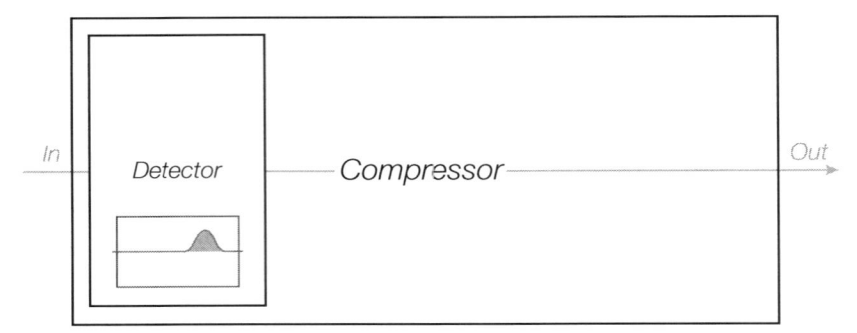

FIGURE 15.39 Wide-band de-esser.

Wide-band de-essers are often used on vocals and other monophonic instruments, while split-band de-essers are used on polyphonic instruments and full mixes to ensure that the remaining frequencies are unaffected. Split-band de-essers are sometimes called dynamic EQs or multiband compressors.

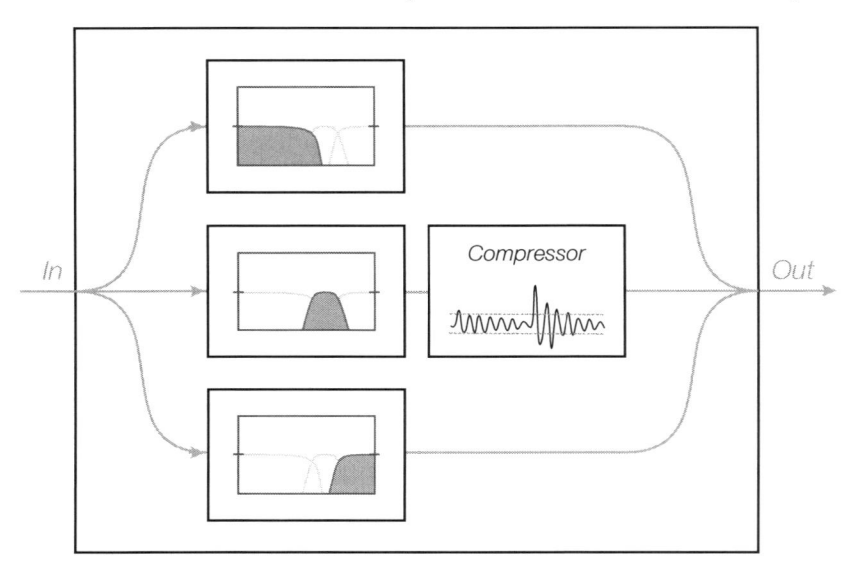

FIGURE 15.40 Split-band de-esser.

15.11 Multiband Compressors

It is possible to split the signal into different bands by using EQ filters and apply one compressor per band. This technique is called multiband compression (which is slightly different to dynamic EQ, depending on the inner working of the processor as explained in Box 15.4) and is very powerful in fine-tuning sounds (Figure 15.41). Multiband compressors can be used as follows:

- When a problem frequency is to be removed dynamically, rather than using a static EQ
- When the dynamic range of a specific frequency area is to be reduced, evening out its presence over time

Examples where multiband compressors can be used include the following:

- Split-band de-essers. As explained earlier, split band de-essers are in fact multiband compressors with simplified controls.
- Adding air to vocals without harshness. Highly compressing the top end of vocals and then raising the level of that band back to its original volume means that the top end (or air) of vocals is not overly dynamic, but always present.
- Making a vocal sound less "nasal." Equalizing 800 Hz out of vocals will probably make them sound too thin, but reducing that band dynamically can resolve the issue.
- Reducing the volume of the kick drum out of a full drum loop by compressing the low end of the signal only.
- Reducing fret noise from fingers sliding along an acoustic guitar.
- Reducing the volume of hi-hats in a full mix.

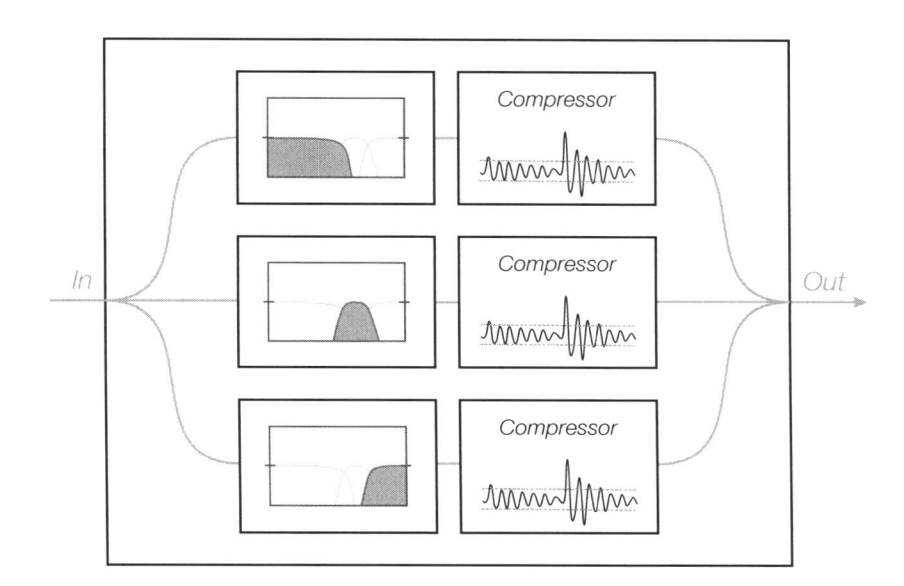

FIGURE 15.41

Multiband compressor.

BOX 15.4

Dynamic EQs are very similar to multiband compressors in what can be achieved sonically. They both can reduce or increase the level of a particular frequency in a dynamic way, according to specific attack, release, and threshold settings. The main difference lies in the user interface and underlying processing. Where a multiband compressor looks and feels like a series of compressors separated by filters, a dynamic EQ looks and feels like a regular EQ (with each band having frequency, gain, and Q controls) with the added feature of being able to control the amount and speed of boosts and cuts depending on input level. At processing level, a dynamic EQ creates less phase shifts as the bands being processed are not separated by filters, but instead the boosts or cuts are dialed-in dynamically by an associated level detector triggered by bandpass filters (Figure B15.1).

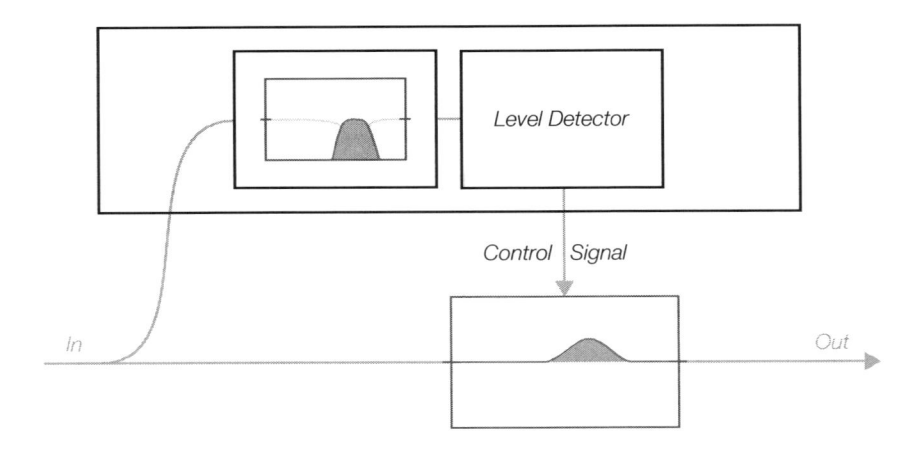

FIGURE B15.1
Dynamic EQ.

15.12 Low-, Mid-, and High-Level Compression

The most common compression settings yield high-level compression: the loudest part of the signal triggers the compressor. With fast attack settings, the peaks are turned down. Transients are often subject to this type of compression, as they are the loudest part of most musical events. High-level compression often affects midrange and high-frequency content, as transients are mostly made of 1–10 kHz (Figure 15.42; Audio 15.36 ▶). Note that even when you are using slower attack settings (and therefore not compressing the high-level signals), most compression settings still fall within the "high-level compression" umbrella. In this case, the frequency content being compressed varies, depending on the input signal.

FIGURE 15.42
High-level compression.

Low-level compression is achieved by using parallel or upward compression (Figures 15.43 and 15.44; Audio 15.37 ▶). This method means that the softest part of the signal is turned up. In this type of compression, the low-level detail is turned up without affecting the transient. Depending on the type of instrument being compressed, low-level compression can affect high-frequency content (where reverb and detail is raised) or low- and mid-frequency content (where the sustain of an instrument is prolonged). It is worth noting that the difference between parallel and upward compression lies in the curve achieved. Parallel compression

FIGURE 15.43
Low-level compression using parallel compression.

FIGURE 15.44
Low-level compression using upward compression.

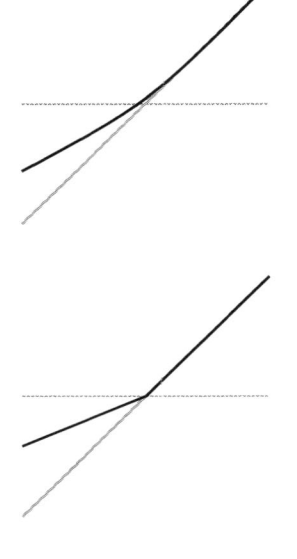

will yield smoother and rounder curve, while upward compression will give linear results.

Mid-level compression can be achieved on compressors that allow for a specific compression curve (Figure 15.45; Audio 15.38 ▶). To achieve this result, you should set the ratio low (1.5:1 to 4:1), set the threshold low enough that the compressor is consistently acting on the loudest part of the signal, and keep the gain-reduction meter consistently moving, indicating that the mid-level signal is kicking in and out of compression. Mid-level compression often affects midrange frequencies and is used to enhance the natural tone of an instrument.

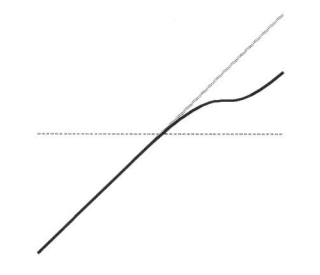

FIGURE 15.45
Mid-level compression.

15.13 Negative Ratios

Some compressors allow for negative ratios to be set. This capability is called cancellation mode, negative compression, or anti-dynamics (Figure 15.46; Audio 15.39 ▶). In this mode of operation, once the signal has reached maximum compression (or full limiting), the volume is further reduced as the input volume is raised. This mode can be used to remove the loudest parts of the signal, produce creative effects, and change the overall groove. When used in parallel, it can allow for further enhancement of the low-level signals.

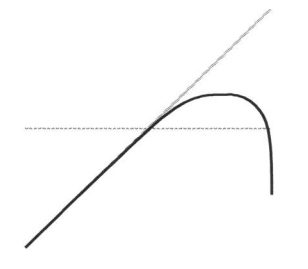

FIGURE 15.46
Negative compression.

15.14 Expanders and Gates

While gates and expanders are mostly used in live sound reinforcement to solve issues such as bleed and feedback, they can also be used as mixing and creative tools in studio situations. These processors work very similarly to compressors, but act on the audio signal below the threshold. They can be used to increase dynamic range and add life to over-compressed signals, as well change the feel and groove of a rhythmic part.

The difference between gates and expanders is quite subtle. A gate (Figure 15.47) reduces the volume of a signal below the threshold by a set value, while an expander (Figure 15.48) reduces the volume at a ratio of the incoming signal.

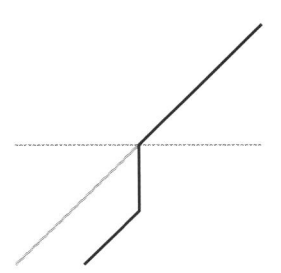

FIGURE 15.47
Level reduction shape for a gate.

FIGURE 15.48
Level reduction shape for an expander.

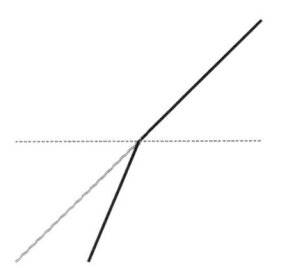

While the controls of an expander/gate are very similar to those on a compressor, it can be useful to redefine them here (Figure 15.49):

15.14.1 Threshold

Similar to that of the compressor, this threshold is the point at which the expander/gate starts or stops working. The difference is that the processing happens when the signal falls below the threshold, where a gate is said to be closed. Once the signal has reached the threshold, a gate is said to be open.

15.14.2 Ratio

The control of the expansion ratio can be thought of as being the same as compression ratio, but reducing the signal below the threshold.

15.14.3 Range

The range of the gate's gain reduction is often set to maximum in a live-sound context, and reduced to a few decibels in a mixing context to keep naturalness. It is the value by which a closed gate is turning the signal down.

15.14.4 Attack

The attack controls how fast the expander/gate opens once the signal has reached the threshold (minus the look-ahead time if enabled). The signal will be "faded in" over the attack duration.

15.14.5 Release

The release controls how fast the expander/gate closes once the signal has fallen below the threshold. The signal will be "faded out" over the release duration.

15.14.6 Hold

The hold controls how long the expander/gate remains fully open after the signal has fallen below the threshold (and before the processor enters its release phase).

15.14.7 Look-Ahead

This control places a delay in the processing chain to allow of the expander/gate to open before the threshold is reached. It is used to avoid clicks if the signal has to go from silence to loud volume in a very short period of time.

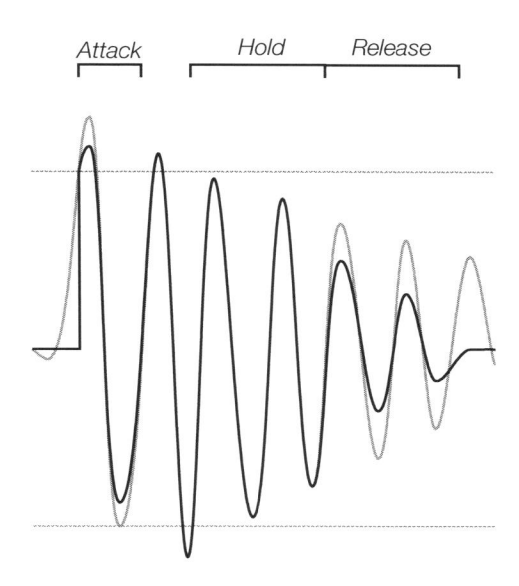

FIGURE 15.49
Gate/expander controls.

15.15 Spill Reduction

In order to reduce spill using a gate, start with a fast attack, hold, and release; full muting through the range setting; and infinite ratio. Lower the threshold until short bursts of sound are coming through for each hit. At this point, raise the hold time to ensure that the full length of the hit comes through, then raise the release time to create a better-sounding fade-out for each hit. You can experiment with different look-ahead settings to ensure that the transients are not being cut, although keeping look-ahead to a minimum can sometimes create more punch, as it may create a slight pop each time the gate opens. If the process sounds a little too unnatural, raising the range can help. If you use a gate to remove spill in a recording, it is important to ensure that the gate is placed before you apply any compression. This precaution is to ensure that the difference between low-level spill and high-level useful signal remains significant enough for the gate to work properly.

While spill is not always an issue and can be used to improve gluing of instruments within the mix, it is important to know when to remove spill and when to leave it in the recording. For a rule of thumb, if you can easily remove the spill by using editing techniques (usually when a track contains more than a few seconds of silence), then you should remove it. The exception exists in sparse mixes where a constant noise appears on one track. In this case, removing the noise in some parts (but not all) may actually make it more noticeable. In order for you to decide whether spill that is not easily editable should be removed or not, it is important to listen to other FX applied on the track and judge whether they negatively impact the "spill sounds." For example, when dealing with a kick-drum track that has snare spill, use a gate on the kick only if the snare is negatively impacted by the EQ or compression placed on the kick.

15.16 Using Gates and Expanders in a Mix

Expanders and gates can be used in mixing to modify the perceived performance and tonal quality of instruments. Generally thinking of these processors as envelope shapers can help you decide where they could be useful in a mix. For example, it is possible to use an expander to reduce the sustain of open hi-hats so that they sound as if the drummer had pressed his or her foot on the pedal halfway through each hit. An expander could also be used on tom-heavy drum rhythms to shorten the length of each hit, therefore resulting in a punchier overall sound without rumble. Using subtle gating, you could make strummed acoustic guitars more aggressive and punchier by making transients louder compared to the sustain of each strum (Audio 15.40 ▶).

Using an expander with fastest attack, release, and hold, you can add distortion to sustained bass instruments where the signal hovers around the threshold. In this case, the expander kicks in and out very fast to creates cuts in the signal. The higher the ratio, the brighter and more obvious is the distortion (Audio 15.41 ▶).

Generally speaking, wherever the transient or the sustain of an instrument needs to be altered, gates and expanders can be used.

15.17 Upward versus Downward

Regular compressors work as downward compressors (Figure 15.50): they reduce audio volume above the set threshold.

Regular expanders work as downward expanders (Figure 15.51): they reduce audio volume below the set threshold.

Upward compression (Figure 15.52) can be achieved by using expanders with ratios lower than 1:1. In this case, the audio volume is increased below the set threshold. Note that this method achieves an effect similar to parallel compression.

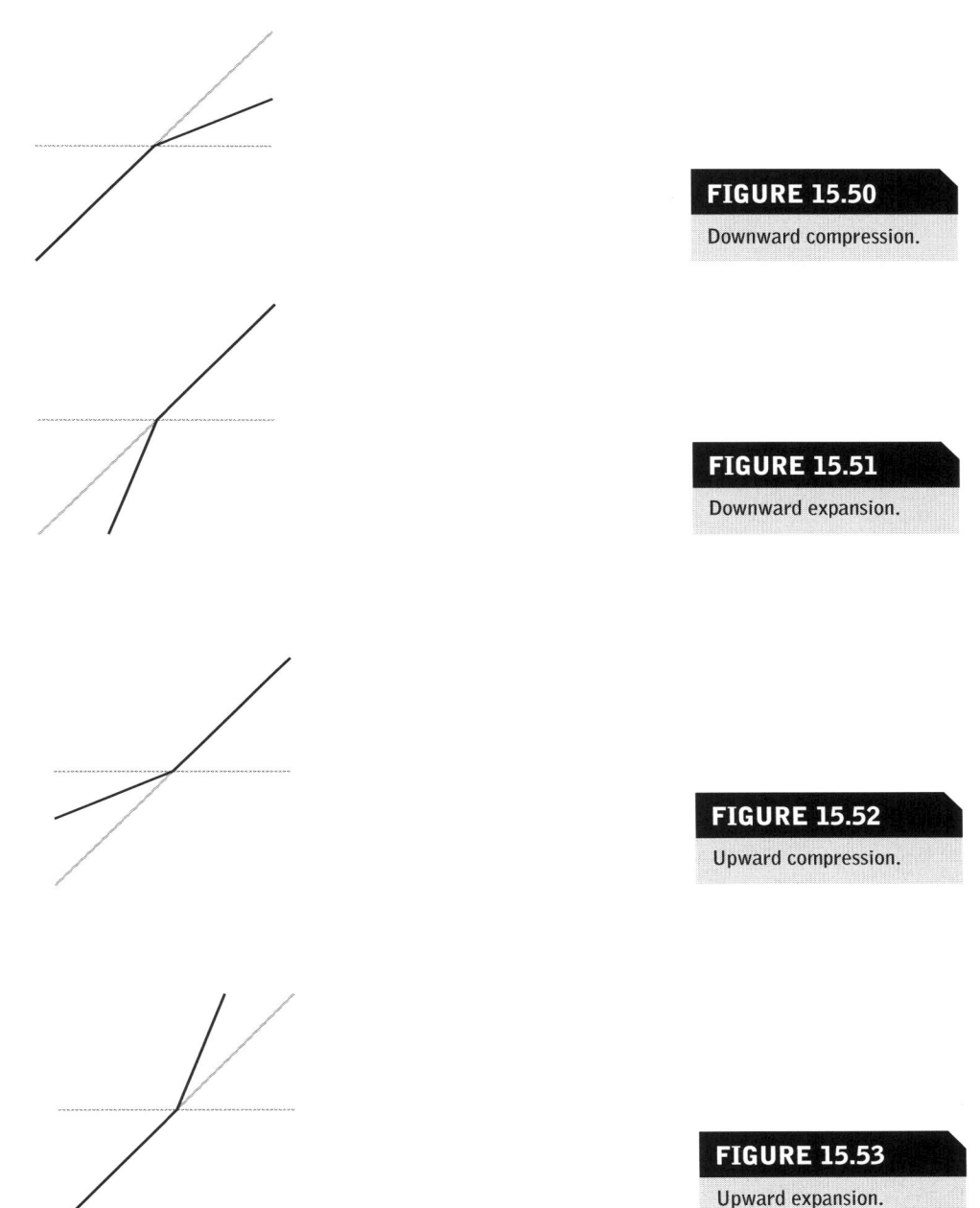

FIGURE 15.50
Downward compression.

FIGURE 15.51
Downward expansion.

FIGURE 15.52
Upward compression.

FIGURE 15.53
Upward expansion.

Upward expansion (Figure 15.53) can be achieved by using compressors with ratios lower than 1:1. In this case, the audio volume is increased above the set threshold. While this type of processing is rare, it sounds similar downward expansion and can bring dynamics back to an over-compressed signal. In the same way that parallel compression achieves an effect similar to upward compression, parallel expansion achieves an effect similar to upward expansion.

Chapter 16

Reverberation and Delays

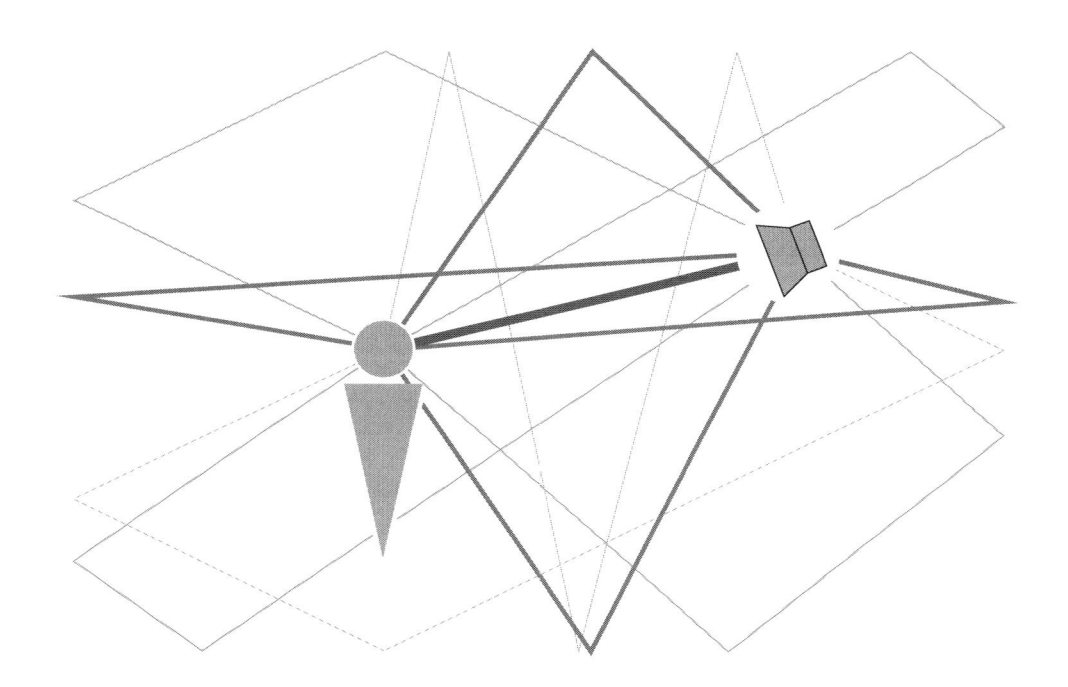

Reverberation (or reverb) is one of the most widely used effects in audio engineering. It is a succession of echoes of the original sound that are so close to one another that we cannot perceive the different occurrences individually.

In nature, these echoes arise from sound bouncing off the walls of a room and reaching the listener's ear after the original sound. Before we look at the uses for reverb, it is important to understand the different types available and the common controls found on reverb processors.

16.1 Reverb Type

16.1.1 Echo Chambers

The earliest form of reverb, the echo chamber (Figure 16.1; Audio 16.1 ▶), was quite simple. Studios used a very reverberant empty room in which they placed a speaker and a microphone (or two if a stereo reverb was to be achieved). Playing back the signal through the speaker and recording the output allowed the engineer to blend the dry signal with the recorded one during the final mix. Digital reverb units often include "chamber" algorithms that emulate this process. An easy way to recreate such effect is to place a speaker in a bathroom and record the signal back from a distance.

FIGURE 16.1
Echo chamber.

16.1.2 Plate Reverbs

Plate reverbs (Audio 16.2 ▶) are large devices (around one to two square meters) that are mainly composed of a thin sheet of metal in which the signal is run. The vibrating sheet then outputs a reverberated version of the signal, which can be blended in with the dry signal. While these devices are very bulky and impractical to use, they can work very well on vocals and backing vocals. The most famous model of plate reverb units is the EMT140 (Figure 16.2). Digital reverb units often include "plate" algorithms that emulate this process.

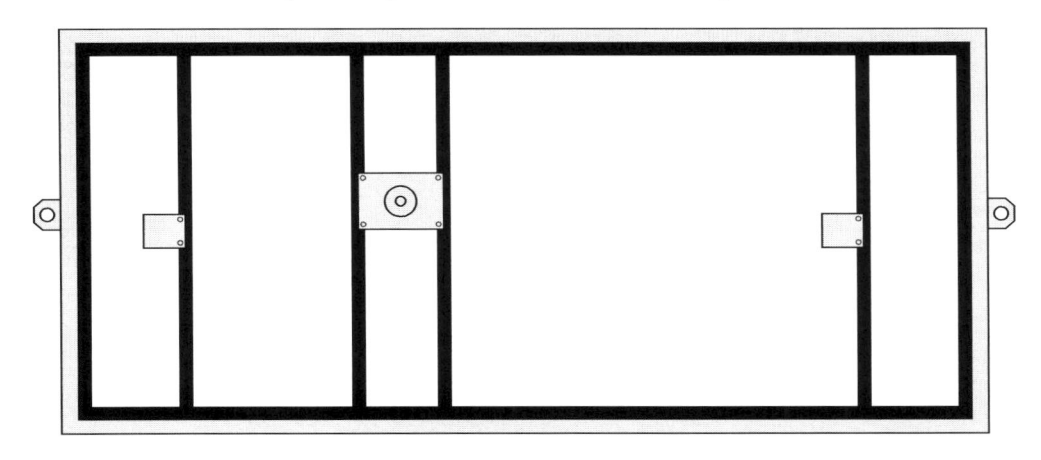

FIGURE 16.2
EMT140 plate reverb.

16.1.3 Spring Reverbs

Spring reverbs (Figure 16.3; Audio 16.3 ▶) work by running the signal through a thin spring (around 20 cm in length). The spring vibrates and its output is a reverberated version of the original signal. Spring reverbs are often found in guitar amps, as they are the cheapest way to create reverb electronically. The general tone of this type of reverb is often said to be "dirty," "cheap," or "vintage," and is often applied to guitars. Once again, digital reverb units often include "spring" algorithms that emulate this process. Note that spring reverbs often emphasize instrument ringing that may otherwise go unnoticed if you are using other types of reverbs.

FIGURE 16.3 Spring reverb.

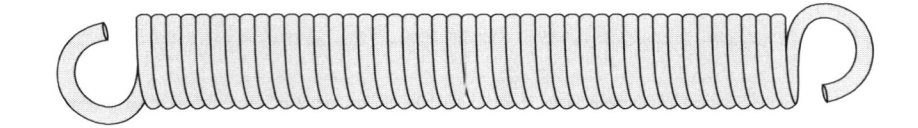

16.1.4 Digital Reverbs

From the mid-1970s onward, digital reverb units started taking over in recording studios. A lot of manufacturers have developed hardware reverb units; some of the most famous models are the EMT 250 (Figure 16.4; Audio 16.4 ▶), AMS RMX16, Lexicon 224, and the TC Electronics System 6000. Because of the high demand for some of these units, plug-in manufacturers have started releasing emulations of the famed reverbs from the 2000s onward.

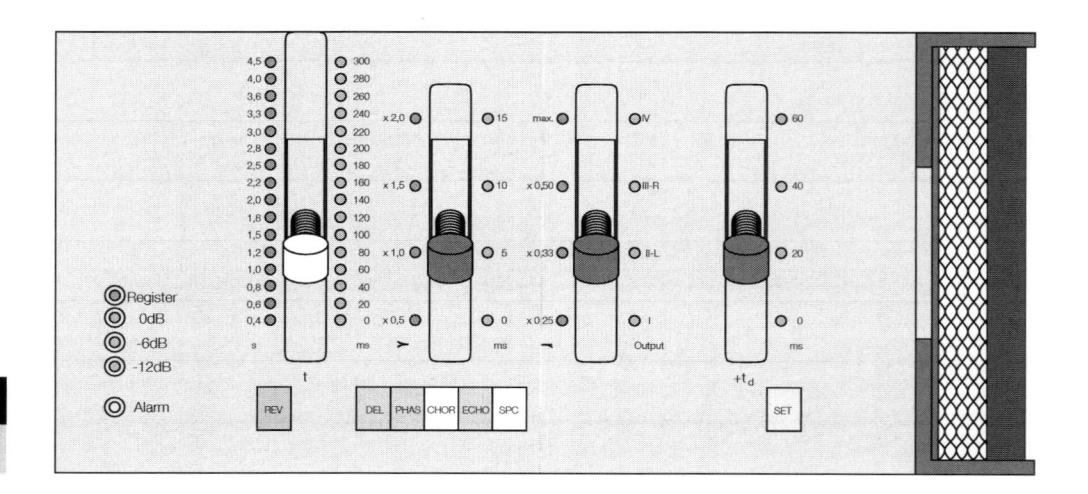

FIGURE 16.4 EMT250 digital reverb.

16.1.5 Nonlinear Reverbs

Nonlinear reverbs are essentially used as effects. While "regular" reverbs decay in a linear way, nonlinear reverbs are used to create unnatural effects such as the infamous gated reverbs (Figure 16.5; Audio 16.5 ▶) of the 1980s or some types of reverse reverbs.

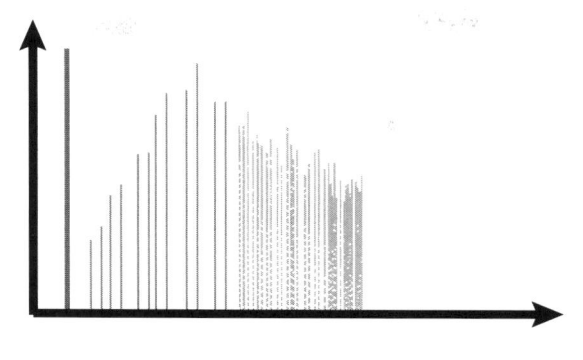

FIGURE 16.5
Nonlinear reverb.

16.1.6 Algorithmic and Convolution Reverbs

In the plug-in world, there are two types of reverb available: algorithmic and convolution. Algorithmic reverbs are mathematical calculations of how a particular room should sound with a given size, geometry, and construction materials, for example. These reverbs can recreate any real or physically impossible acoustic space. Advancements in computing powers made convolution reverb possible around the turn of the twenty-first century. Convolution reverbs with presets (Figure 16.6; Audio 16.6 ▶) record the impulse response of a particular room by playing a short burst of sound or sine sweep in a space, and recording the reverb that it produces as it excites the room. This impulse response is then available in the plug-in host, which allows a signal to be run through and have similar reverb qualities be applied to it. In theory, the reverb is exactly the same as if the signal were being played in the original space. Most impulse responses feature different distances between the sound source and the listener, allowing for different reverb sizes to be applied.

While convolution reverbs tend to sound more natural than algorithmic reverbs, neither is better and both types have their place in modern productions.

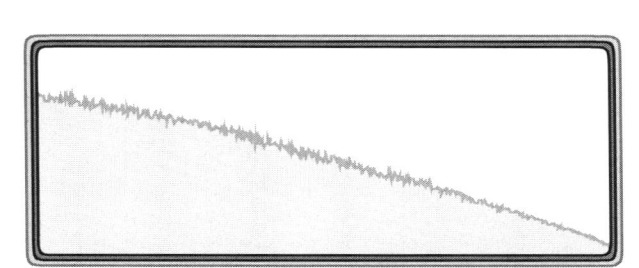

FIGURE 16.6
Convolution reverb preset.

It is possible to record your own impulse responses when you find yourself in an interesting space. There are two methods for doing so: recording a sine sweep and deconvolving it later in the plug-in, or recording a transient short burst of broadband sound and loading it into the convolution reverb. The second method is a lot easier and faster to implement in a music production setting. Without going into the nuts and bolts of convolution technology here, we can explain that recording a short burst of sound such as a clapper (used on film sets), a balloon popping, or a starter gun in a space allows for the sound to be loaded into the reverb and be able to then run different instruments through that space. While this method is often used during film location recording to allow for the same reverb to be applied to automated dialogue replacement (ADR) and Foley as what is recorded on set, in a music production context, it can be beneficial to record the impulse response of a room and apply the recording to the different instrument in the mix to glue them together. For example, if the drums were recorded in a good live room, and everything else was done quite dry, it is possible to "place" the dry instruments in the same live room and therefore create coherence between the instruments.

16.2 Early Reflections/Reverb Tail

As the name suggests, early reflections (or ERs) refer to the reflections of the original signal being heard first and are the first reflections off the room's surfaces. ER echoes are more spaced in time from one another than in the reverb tail, and the spacing allows our brain to interpret the size, geometry, and construction materials of the room. The reverb tail happens after the ERs have been heard (Figure 16.7; Audio 16.7 ▶). The echoes of the signal are so close to one another during the reverb tail that we perceive them as being a single "sustained" echo of the signal.

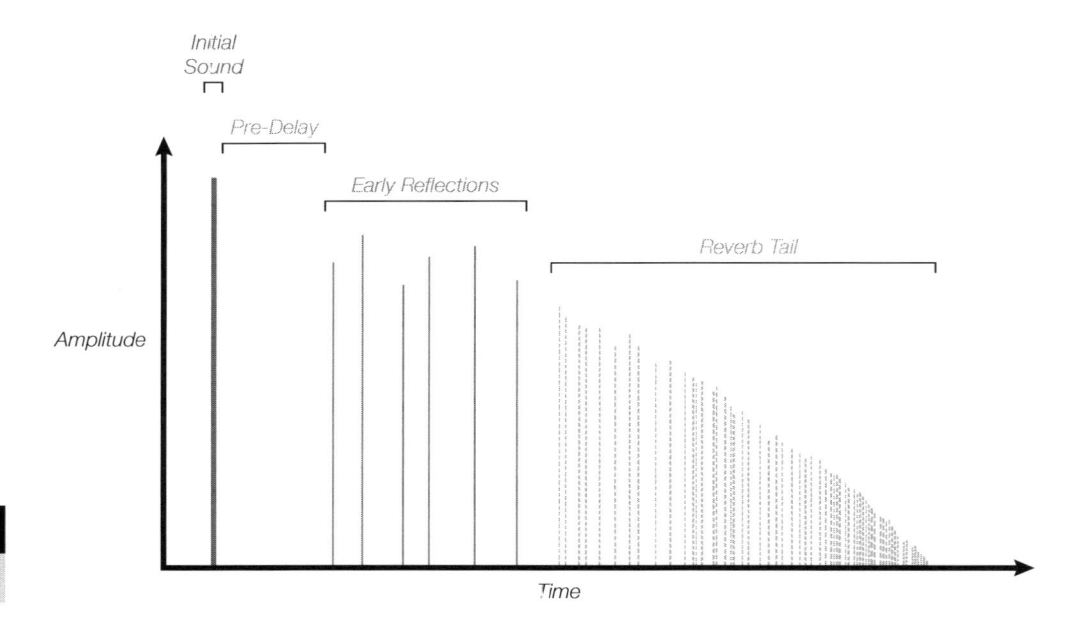

FIGURE 16.7 Reverberation parts.

The correct way to set up a reverb largely depends on the intended purpose. There are three main reasons for using reverb in a mix:

- To place an instrument within a specific virtual room
- To enhance the tone of an instrument
- To push an instrument to the back of the virtual soundstage

16.3 Placement Reverb

Instruments in a mix often need to sound as if they belonged to the same band and were playing in the same acoustic space. The vision for the song should dictate whether the band will be placed in a small room, a church, or a concert arena, for example. Mixes from audio engineering students often sound very confusing and disjointed due to the misuse of reverbs. If a dry snare is mixed with overheads recorded with a rock band in a very live room, then a plate reverb is placed on vocals, and guitars are sent to church-type reverb, the listeners will be confused as they try to picture this band into the virtual acoustic space. Using the same reverb for every instrument will ensure that a cohesive virtual space is created.

Since the early reflections give information on the size and construction of the space, only the ER part of the reverb should be used for placement. Most reverb units have separate controls for ER and tail, and that separation means that the tail can be muted on the reverb return used for placement. The quality of the reverb is often important, as high-quality units will give more natural results.

To ensure that you are applying this technique correctly, it is good practice to set up and use only one return for ERs (Figure 16.8). Sending any instrument that needs to be placed within the chosen acoustic space to this return will give the impression that all instruments are playing in the same room, reinforcing the sense of togetherness.

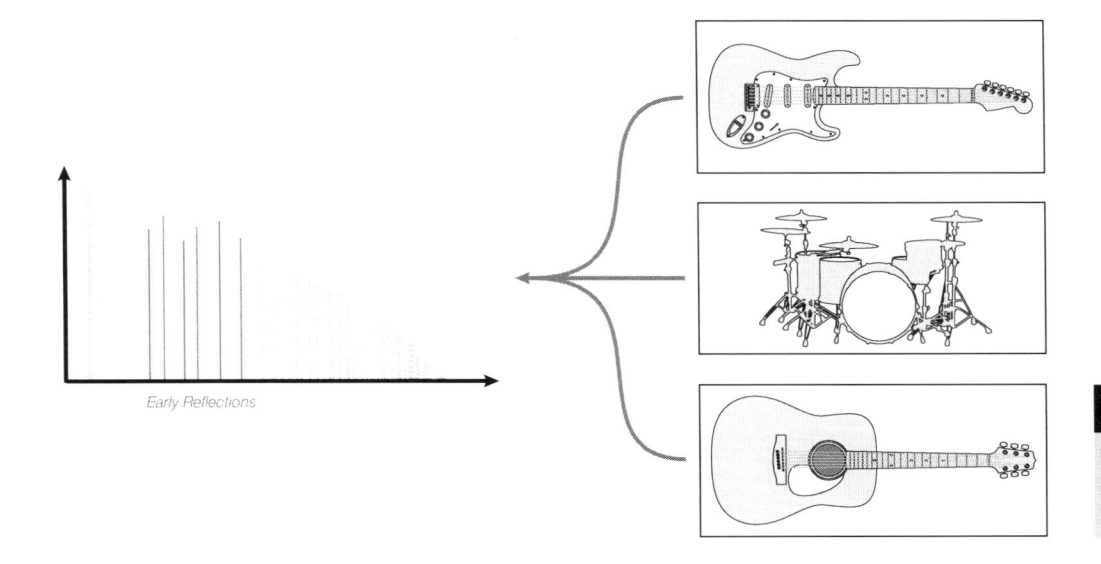

FIGURE 16.8

Single-reverb early-reflections return for all instruments in the mix.

16.4 Enhancement Reverb

Reverb can also be used to enhance the tone of an instrument; most instruments sound better with reverb added. You should be careful when adding reverb to embellish an instrument, as too much reverb can ruin a mix by blurring the lines between instruments and muddying the low end. Because different reverbs have their own particular sounds, the tonal qualities of each unit can be used to enhance the different sounds in the mix. For example, it is not uncommon to set up short, medium, and long reverb returns and send the necessary instruments to them. Since the EMT250 is great at enhancing lead vocal tones, this processor can also be set up here. The same could be done with an emulation of the EMT140 on backing vocals, for example.

The part of the reverb that is used for enhancement is the tail. All of the different returns used for enhancement should therefore use only the tail part of the unit and early reflections should be disabled (Figure 16.9).

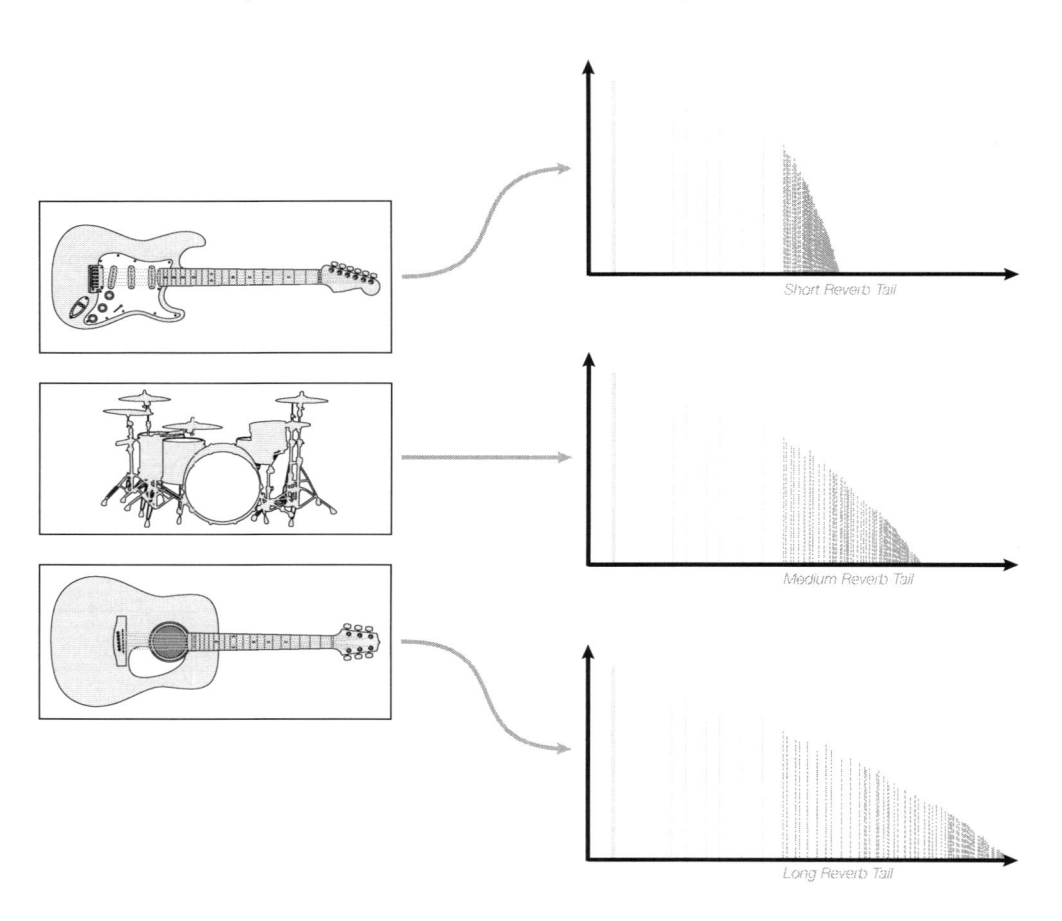

FIGURE 16.9
Multiple reverb tail returns for different instruments in the mix.

Since the aim of enhancement reverb is usually to thicken the sound or add depth, alternative processors can be used to achieve one or the other. For thickening, effects such as chorus, doubling, and slap delay can achieve similar results, as they will also blur the instrument just as reverb would. For depth, effects such as EQ and delay can help.

16.5 Reverb for Depth

In some cases, reverb needs to be used to push an instrument to the back of the mix. Applying reverb to any instrument will automatically push it back without the need to reduce its volume. Here are some tips that can be used to push an instrument to the back of the mix, or, on the contrary, bring it forward:

16.5.1 To Push Backwards

- ERs must be as loud or louder than the dry sound. As more ERs and less direct sound are being heard, the instrument is pushed further away from the listener, while staying in the same virtual space.
- Use longer reverb tails. Long tails are synonymous with large acoustic spaces and often give the illusion that the sound source is far away from the listener.
- Sound source and reverb return should be mono. In nature, as a sound moves away from the listener, it is perceived as coming from one point. Any sound source that is farther than a few meters away from the listener is effectively heard in mono.
- Equalize high frequencies out of the source. Sounds that are farther away tend to lose their high-frequency content. Losing some of the detail and brilliance of a sound will help in pushing it back in the mix. Note that this method should be applied to the original signal before it reaches the reverb. Extra equalizing can also be done after the reverb to push the instrument even further back.
- In some cases, using all-pass filters to turn high frequencies out of phase can create the illusion of distance, as high frequencies get absorbed more than low frequencies with distance.

16.5.2 To Bring Forward

- While early reflections can be necessary to place and instrument within a specific environment, using less ER and more dry sound will ensure that the instrument stays at the front of the mix.
- Do not use reverb tails if you want to achieve closeness. For opposite reasons than those listed before, long reverb tails are often synonymous with large spaces and should therefore be avoided.

- Sound source and reverb return should be stereo. As previously mentioned, mono returns push the sound back; therefore, using stereo returns will achieve the opposite effect.
- Add low and high frequencies to the sound source. Adding bass will emulate the proximity effect that most listeners are accustomed to, while adding treble will bring out the detail of a sound and therefore bring it forward in the mix. Once again, using EQs (and sometimes harmonic enhancers) before the reverb will give the best results. Extra equalizing can also be done after the reverb to bring it forward even more.

16.6 Reverb Settings

For either of these applications, it can be hard to pick the right reverb settings for the song being worked on. The following guidelines can be useful when you are choosing the type of reverb being used.

16.6.1 Reverb Length

Generally, short reverb times are used for songs with fast pace (Audio 16.8 ▶). This use ensures that the reverb is only filling small gaps between instruments and not introducing frequency masking or blurring the separation between them. For songs with slower pace, longer reverb can be used to fill more of those gaps between instruments and ensure that a "backdrop" is created behind the elements of the mix.

16.6.2 Reverb Thickness

The choice of using very rich- or dry-sounding reverbs, whether through the use of different presets, "damping/absorption" control (emulating the construction materials of the elements in the virtual space being recreated), or "diffusion/density" control (the number and spacing of reflections in the virtual space being recreated), will depend on the instrument being reverberated and the overall frequency content of the mix (Audio 16.9 ▶). For example, if a mix already contains a lot of instruments that are filling a lot of the frequency spectrum, using rich- and lush-sounding reverbs on everything may create a mess. In this case, using mostly dry reverbs in the mix and picking only a couple of instruments to feature a lusher reverb will be more appropriate. Another example is using reverb as a tonal enhancement tool, just as an EQ would, on an instrument that needs to feel larger. In this case, a lush reverb can very effectively create the required effect without the need for such enhancements as EQ, saturation, and compression.

16.6.3 Reverb Quality

The choice of using natural- or unnatural-sounding reverbs depends on the type of song, instrument, and feel required (Audio 16.10 ▶). It is commonly accepted that acoustic instruments benefit from realistic and natural-sounding reverbs, while amplified instruments can use more exotic and unnatural ones. The exception should be made when trying to fit amplified instruments or synthesizers within a mostly acoustic song. In this case, using a more natural-sounding reverb can help blend everything together.

16.7 Using Convolution Reverbs

Convolution reverbs can be used differently from the way their algorithmic counterparts are. If you are using convolution reverbs for their naturalness, set up different returns with the impulse response of the same acoustic space recorded at different distances. All of the instruments can then be sent to the different returns, depending on how far back they are supposed to sit in the mix. Note that it is preferable to send all of the instruments to the reverbs at the same volume for this technique to work properly. This technique is very effective when you are dealing with acoustic ensembles such as orchestras or folk groups (Figure 16.10; Audio 16.11 ▶).

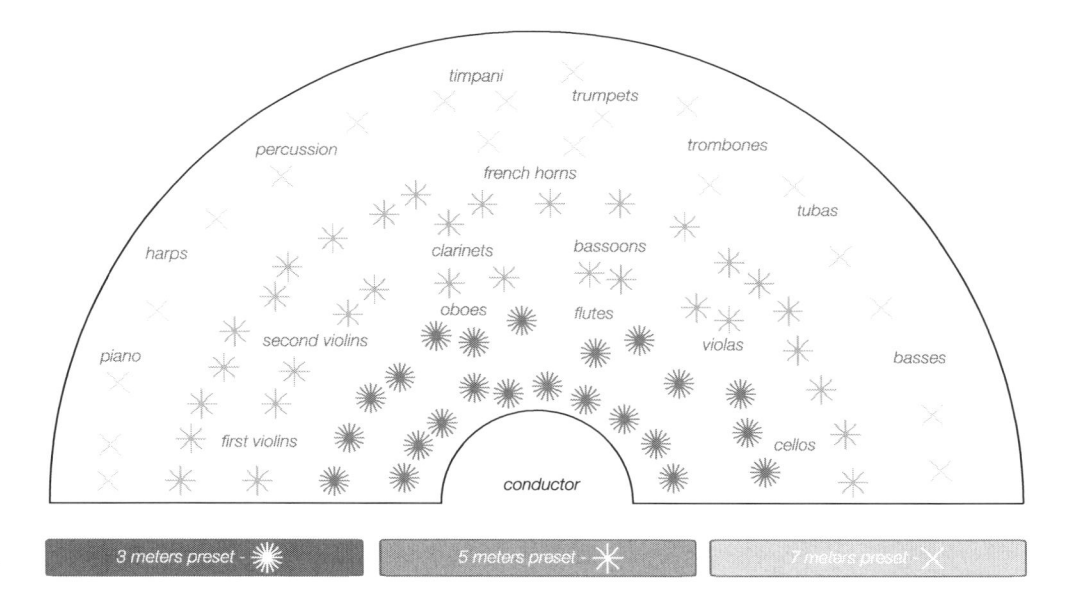

FIGURE 16.10
Using different convolution reverb presets with orchestral recordings.

16.8 Reverb Tips

16.8.1 Equalizing the Reverb

Equalization should be applied before the reverb unit, that is, pre-reverb. Because sounds do not reach reflective surfaces with excessive low-end energy in the real world, an HPF should be applied to the signal reaching the reverb. This step will help clean up the low end of the mix. Similarly, it is advisable to apply an LPF before the reverb to ensure that the reverb never feels overly harsh. It is often preferable to cut some of the main frequencies of the instrument being sent to ensure that the dry signal cuts through unmasked by the reverberation (Figure 16.11; Audio 16.12 ▶). Very clean EQs should be used for this process, as using an EQ with too much color can affect the sound of the reverb in unpredictable ways.

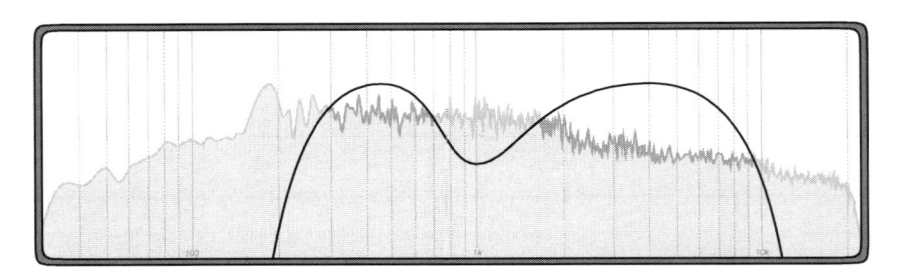

FIGURE 16.11
Vocal reverb EQ (pre-reverb).

16.8.2 Compressing the Reverb

Another trick to make the reverb blend better with the dry signal is to use compression on the signal reaching the reverb (Audio 16.13 ▶). The aim here is to completely cut the transients reaching the reverb so that only the sustained part of the signal gets affected. This step will help keep the original transients dry, while gluing the dry sustain with the reverberated sustain. To that end, using a healthy amount of compression is necessary, and while a fast attack setting is needed, the compression should still be rather smooth.

16.8.3 Lush Reverb

The best-sounding halls in the world often exhibit the same reverb characteristic: the low-end frequencies decay more slowly than treble. This phenomenon can be used to enhance the tone of the reverb (Audio 16.14 ▶). Some units have dedicated controls over the decay time of the different frequencies, while others have only one control for the decay. For reverbs that do not have individual controls for the decay of specific frequency ranges, using multiband compressors and transient designers after the reverb can help bring out the low end of the reverb.

Another trick to make the reverb sound warmer is to slightly saturate it using by soft clippers or other smooth-sounding distortion processors. The essential point here is to be very subtle, but the harmonics added can add thickness to a reverb return.

16.8.4 Reverb-Tail Modulation

Some instruments can benefit from gentle modulation on the reverb tail. Adding such processors as chorus, flanger, and tremolo can work wonders in thickening fretless instruments that are always "approximating" pitch, such as strings, vocals, or trombone. Be careful when you are using this technique, as instruments that have a constant fixed pitch every time a note is played (piano and guitar, for example) do not lend themselves well to modulation.

16.8.5 Using Pre-delay

Pre-delay—placing a short delay before the reverb—will allow some of the original dry signal to be heard before any reverb is added. Pre-delay can be quite useful for transient heavy material such as drums, as it keeps the dry punch of each hit before the reverb is heard (Figure 16.12; Audio 16.15 ⏵). This technique is most often used on reverb-tail returns, and the delay should not be more than 50 ms; otherwise, the reverb would be perceived as an echo. Setting the pre-delay to a musical value can also help the reverb move "in time" with the music. If the pre-delay time is set too long, using a transient designer to reduce the attack of the drum before the reverb can help in reducing the illusion that two separate hits are played.

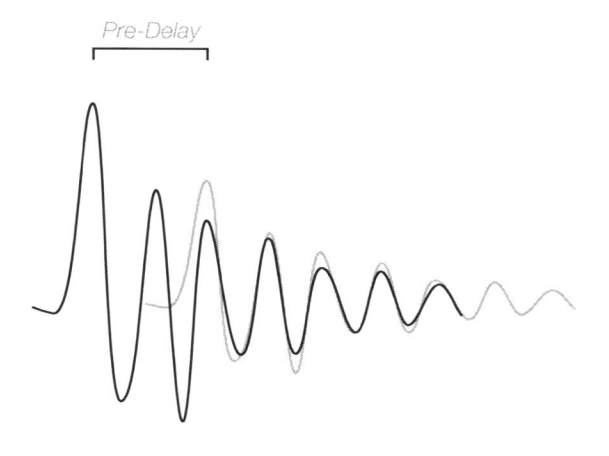

FIGURE 16.12

Reverb pre-delay.

16.8.6 Side-Chain Compression on Reverb Tails

In much the same way that adding pre-delay can help drums keep their punch, turning the reverb down when the dry signal is playing will ensure that the detail and definition of the original sound are not lost. This technique is very useful on vocals, as the reverb tail from one word can mask the next. Using a compressor inserted after the reverb with its side-chain input set to receive the dry vocals is the easiest way to implement this technique. This technique works even better if the original vocal triggering the side chain has been highly compressed, allowing the compressor to react to every word instead of only the loudest ones (Figure 16.13; Audio 16.16 ⏵). While this method is very useful for reverb tails, ducking ERs can cause inconsistencies in the virtual acoustic space created.

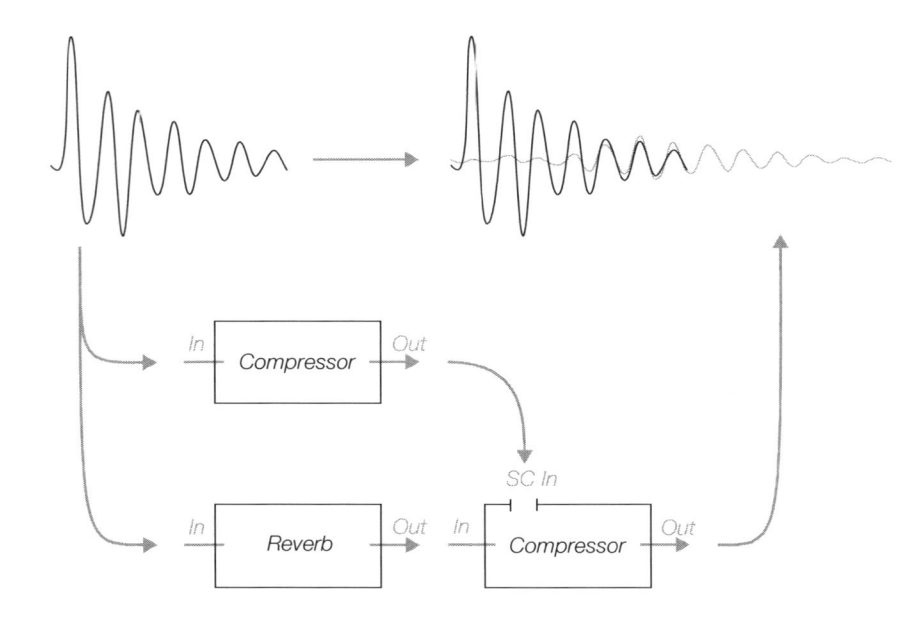

FIGURE 16.13
Ducking reverb return from compressed dry signal with a compressor.

16.8.7 Uncompressed Reverb Send

Sending an instrument to the reverb before using compression on the dry signal can help keep the instrument's dynamics (on the reverb return), while controlling the dry signal. This technique is often used on acoustic guitars and other acoustic instruments (Figure 16.14; Audio 16.17 ▶).

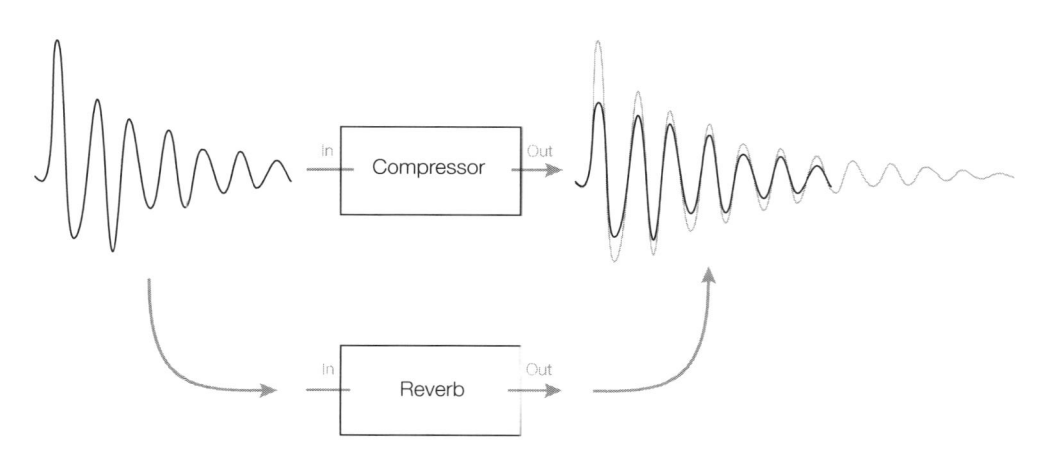

FIGURE 16.14
Sending uncompressed signal to reverb.

16.8.8 Reverb Transmission

When you are applying a long reverb on one instrument, the illusion that other sounds in the same frequency area are also reverberated occurs. For example, adding a long reverb on an acoustic guitar will give the impression that some of the drums are reverberated as well. Because of this phenomenon, reverb should never be added when instruments are soloed; instead, all instruments should be playing together when you are judging reverb levels.

16.8.9 Is Reverb Needed?

It is always too easy to add reverb to elements of a mix to add such elements as coherence, glue, and sweetness. A question that should be asked at the start of any recording project is whether the natural ambience of the recording space could be enough not to have to use extra reverb in the mix. If this approach is possible, use it. If all the instruments in the song are recorded at the same studio, careful microphone placement should allow for the recording to have the same overall tone printed on everything and should negate the need for extra reverb. The end result will be a cleaner and punchier mix.

16.9 Delays

Delays can sometimes replace reverbs in dense mixes where adding depth to an instrument without the blur associated with reverbs is required. In this case, a few settings need to be taken into consideration in order to properly blend the delays with the rest of the mix. First, the amount of feedback or number of echoes must be decided. The more echoes present, the more the delay will be obvious to the listener. If multiple echoes are used, then the delay time should be set in sync with the tempo of the song to ensure that the delay does not distract the listener. If a single echo is used, then the timing can be set more arbitrarily, depending on what sounds appropriate for the mix. Generally speaking, the shorter the delay time, the less important it is to have it synced to the tempo. Note that synced delays tend to glue instruments together a little better. Most delays will also have an LPF built in the feedback line. This technique has two uses in music production: emulating tape-delay effects, and ensuring that the delay effect is not distracting if multiple echoes are used. This precaution is particularly important if the mix is quite dense, as bright echoes can often clutter the top end of the mix. Another tool used to blend delays better with the mix is to try to reverse the polarity of the delay return. While this reversal does not always work in your favor, it can sometimes help glue the delay and dry signals better.

When the delay time is set to a musical value of the song's tempo, subtle delays to hi-hats or snare tracks can emulate how a drummer would play ghost notes in between the main hits. Automating the volume send to this delay can then add a "human" feel to those added hits by giving them more or less volume. This delay can also be slightly faster or slower to give a rushed or relaxed feel to the performance. When we take this concept further, an LFO can modulate the delay time by a very small amount to move the ghost notes around and give an even more realistic performance. The performance could even be completely changed by using dotted notes or triplet delay times.

When you are using delays on vocals, it is often useful to automate the send to the delay to repeat only the last word or syllable being sung. Echoes on a full vocal phrase can clutter a mix and the musical effect on the listener is often no better than simply adding echo to the end of a phrase. If the full vocal line needs to have echo, using compression on the delay return side chained from the dry voice can ensure that the echoes appear only during gaps between words rather than overlapping with the singer.

As mentioned in Chapter 12, "Mixing," instruments must be balanced in the stereo field by frequency content. Thus a shaker panned to the left must be counterbalanced by another high-frequency instrument on the right. If no other instrument is available for this task, sending the shaker to a delay on the right side could counterbalance frequencies without the need for another instrument.

Extra effects can be added after a delay to further process instruments. For example, adding saturation or harmonic enhancers can make the delay sound warmer. Using compression can help with the consistency of the echoes. Using a reverb can add another level of depth and create extra-long tails. Using transient shapers to extend the sustain can add more body to the echoes.

Delays can also create all sorts of well-known effects, depending on their time, modulation, and other parameters available in any DAW's simple delay. It is important to note the two different categories of effects available when you are using delays: fusion effects and echo effects. Fusion effects are created when the delays are so short in time that they seem to fuse with the original sound to create a new composite sound. Echo effects, on the other hand, are created when the delays are far enough apart that our brains can process them as two different audio events: the original and the delayed sound. It is generally accepted that the changeover between those effects happens around 30 ms, although this switch can happen earlier for short percussive sounds and later for long swelling sounds.

16.9.1 Fusion Effects

16.9.1.1 Comb Filter

The simplest form of effect that can be created with a single delay is comb filtering. This effect is created by adding a single delay of up to 10 ms to a sound source. The phase cancellation between the original and delayed sounds creates peaks and dips in the frequency spectrum that are harmonically related. This effect results in a "comb-shaped" EQ applied to the sound (Audio 16.18 ▶).

Note that if the original and delayed sounds are panned on opposite sides, phase stereo is achieved (for more on this topic, see Chapter 13, "Panning").

16.9.1.2 Flanger

The well-known flanger effect can be obtained by slowly modulating the time of a single delay of up to 10 ms with an LFO. This method effectively creates a moving comb filter. While the original tape flanging was achieved by duplicating the signal onto a second tape and slowing down this copy by hand, more "classic" flanging effects can be achieved by adding feedback, thus accentuating the comb filtering (Audio 16.19 ▶). This effect can also create "fake stereo" by duplicating the original signal, hard-panning both copies left and right, and setting the flanger on each side slightly differently

See Box 16.1 for further discussion of flangers and phasers.

> **BOX 16.1**
>
> *Even though flangers and phasers often sound very similar, they differ in their inner working. Phaser effects are achieved by slowly modulating the phase of a series of all pass filters with an LFO. This method effectively creates a series of nonharmonically related notch filters that move equally across the spectrum. Once again, feedback can be used to accentuate this effect (Audio B16.1 ▶).*

16.9.1.3 Chorus

Chorus effects are essentially the same as flangers but the delays vary between 10 ms and 20 ms and very little feedback is used. An LFO is also slowly modulating this delay's time, but since the delays are longer, the comb-filtering effect is softer (Audio 16.20 ▶). To create a thickening effect on mono instruments, you can often send the instruments to a stereo delay where the left and right sides have slightly different delay times and LFO modulation rates.

For information on doubling effects, see Box 16.2.

> **BOX 16.2**
>
> *Using slightly longer delay times (10 ms to 30 ms) and detuning the copy by a few cents creates a doubling effect. This effect aims to emulate how an overdubbed instrument would sound and sounds similar to micro-shifters, where the LFO is modulating the detuning rather than the delay time (Audio B16.2 ▶).*

16.9.2 Echo Effects

16.9.2.1 Flam
A flam delay is a single repetition of the sound delayed by 30 ms to 50 ms, which gives the impression of a double hit (Audio 16.21 ▶). This effect is similar to a drummer hitting the snare with two sticks as once.

16.9.2.2 Slapback Delay
A slapback delay (Box 16.3) is similar to a flam delay (single repetition of a sound) but the delay time varies between 50 ms and 200 ms. This effect can be identified as an intentional delay rather than a double hit (Audio 16.22 ▶).

> **BOX 16.3**
>
> *In some modern genres of music, using slapback delays instead of reverb can give instruments a sense of "wetness" without the added blur that reverbs introduce. This technique is often used by producers trying to fit vocals within busy mixes. Since a slapback delay is a single delayed copy of the original sound, using this process instead of a long reverb tail can free up valuable space in the mix. Depending on the type of vocals, using a musical value for the delay time can be beneficial or, on the contrary, detrimental to the sound. For a rule of thumb, musical values should be used for very "rhythmic" vocal performances, while nonmusical values should be preferred for more general uses.*

16.9.2.3 Tape Delay
Tape delays are a type of echo effect that aims to emulate how a signal would degrade with each pass through a tape machine. Each resulting echo of the original sound would be slightly degraded (high-frequency loss) from the previous echo (Audio 16.23 ▶). Note that tape delay times originally varied by moving the playback and record heads closer or further away from each other, allowing the user to achieve a range of different delay times.

16.9.2.4 Ping-Pong Delay
A ping-pong delay creates echoes of the sound alternating between the two speakers (Audio 16.24 ▶). It can be created by sending the original signal to a delay with its output set to one speaker only. A copy of that delayed signal is then sent to a second delay (with the same time settings) with its output set to the other speaker. A copy of this second delay is then sent back to the first delay. The level of this send controls the overall feedback amount.

16.9.2.5 Multi-Tap Delay
A multi-tap delay can be used to create specific echo rhythmic patterns. These processors use several delay lines each set to a different time to create more complex rhythms (Audio 16.25 ▶).

Chapter **17**

Saturation

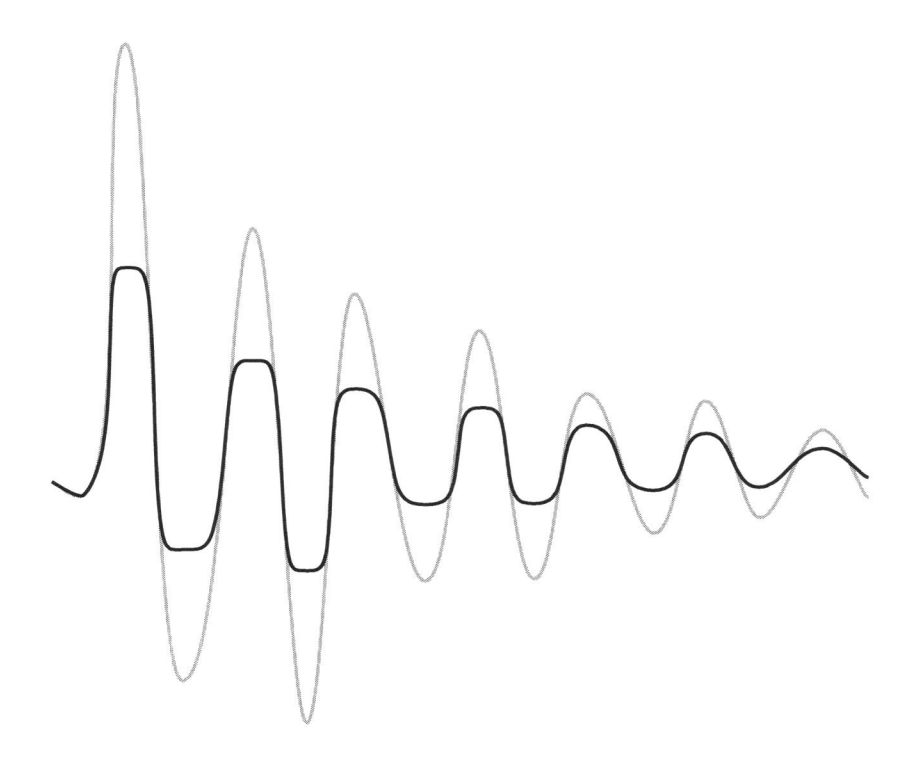

Since DAWs and plug-ins have now almost completely replaced tape machines and analog processors in recording studios, the need for saturation is getting more and more important in order to inject analog-sounding characteristics to otherwise cold-sounding digital recordings. Most commercial productions now have some sort of saturation on each track of the mix. Saturation can be achieved through a range of different processes, all of which have their own characteristics.

17.1 Hard Clipping

Hard clippers give fuzzy distortion, which we mostly consider being "digital" distortion. They work by simply by squaring off the top of each waveform when the clippers are driven (Figure 17.1; Audio 17.1 ▶). Hard clippers are most appropriate for modern electronic music and to raise the average volume of drums.

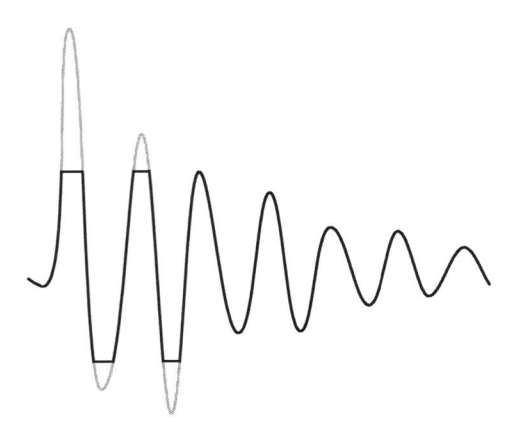

FIGURE 17.1
Hard clipping.

17.2 Soft Clipping

Soft clippers are very similar to hard clippers in the way that the top of the waveform is squared off, but the clipping curve is smooth, rather than abrupt (Figure 17.2; Audio 17.2 ▶). Soft clippers sound warmer and more analog than their hard counterparts. They are often used on sub-bass instruments to extend their presence higher in the frequency spectrum. Soft clipping can be thought of as being similar in process to tape saturation without the tonal character associated with tape.

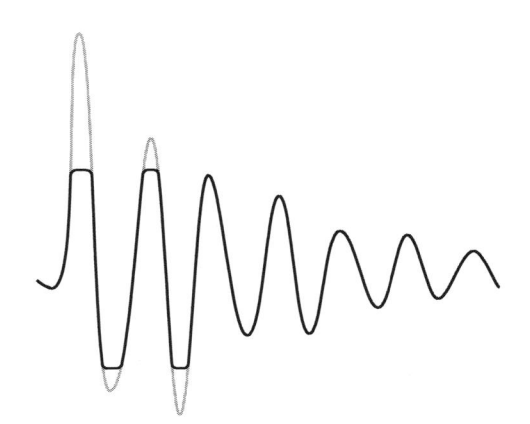

FIGURE 17.2
Soft clipping.

17.3 Limiters

Limiters with a hard knee and fast release can give similar results to those of hard clippers, while those limiters with soft knee and slow release will sound closer to soft clippers (Audio 17.3 ▶). Experimentation is key in figuring out whether a dedicated hard/soft clipper or a limiter sounds better on a particular instrument.

17.4 Summing Console Emulations

A lot of mixing-console-emulation plug-ins have been developed since the early 2000s. All of them work in similar ways: they add nonlinear saturation to the incoming signal in a similar way that driving a console would (Audio 17.4 ▶). Some of the more advanced emulations also imitate the crosstalk that happens between the channels of a mixing console. These plug-ins can be useful in adding nonlinearities between audio tracks.

17.5 Analog Modeled Plug-ins

Some plug-ins model analog equipment and can be overdriven to achieve similar results. For example, using a modeled EQ with all the settings flat, but sending signal to it more hotly than its nominal operating level, will result in the unit behaving as an analog saturation device.

17.6 Tape Saturation

In much the same way that console emulations introduce saturation, tape emulations also alter the harmonic content of a signal when they are overdriven (Audio 17.5 ▶). The tone of the saturation can also be controlled by changing, for example, the type of tape, machine, speed, and flux.

17.7 Tube Saturation

Tubes (Figure 17.3) can be used to introduce extra harmonics in the signal and can be wired as triodes or pentodes (Audio 17.6 ▶). While experimentation is often necessary to obtain the best results, triodes often have a darker quality, and pentodes are often brighter.

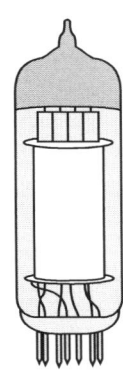

FIGURE 17.3
Vacuum tube.

17.8 Harmonic Enhancers

These processors can work in very different ways. Some synthesize extra harmonic content, while others clip the signal and equalize the result dynamically. These processors can be used as a means to add warmth to instruments in the mix. EQs are different from harmonic enhancers, as EQs can boost only frequencies that are present in the signal, whereas enhancers can create new harmonics based on the content already present. In essence, EQs shape sounds, while enhancers change sounds. There are many uses for enhancers in a mixing context, such as decorrelating channels (changing the frequency content of left and right channels, or even mid- and side channels), ensuring that mid- and side channels have a similar "feel" by making their frequency contents similar (see Chapter 18, "Mastering," for more on this topic), and generally changing an instrument's sound to create thicker tones (Audio 17.7 ▶).

17.9 Parallel Processing

If the aim of the saturation is to add character to a signal, it can often be useful to run the procession in parallel rather than directly on the signal. This process allows for finer control over the saturation characteristics. Depending on the type of saturation being used, two ways of blending the effect can be used (Audio 17.8 ▶):

- Very little saturation in the parallel chain, but mixed in quite high. This method will slightly color the sound without overpowering the clean signal.
- Highly saturated parallel chain, but mixed in quite low. This method allows for high-saturation characteristics to be heard, but blended in quite low as not to overpower the clean signal.

Chapter 18

Mastering

Mastering is the last process before the record is ready for release. While the definition of the term varies from engineer to engineer, there has been a clear shift since the turn of the twenty-first century in what is required from a mastering engineer. Mastering engineers used to be responsible for tasks such as ensuring that masters could fit properly on vinyl, that CD data was entered properly, or International Standard Recording Code (ISRC) were embedded within the files. The decline in CD and LP sales, as well as the rise of affordable home recording equipment, has turned mastering into a "quality control" process rather than the original transfer of mediums. With bedroom producers on the rise and more songs now being mixed on headphones or less than ideal monitoring environments, there is a need for such quality control before a recording is released. These days, mastering is more about getting a fresh pair of (professional) ears to listen to the mix and make the necessary adjustments for it to compete on the commercial market. Sometimes mastering means doing nothing at all. Sometimes it means fixing enormous problems. Such is the nature of the job. Choosing mastering engineers should be based on the quality of their hearing, their monitoring environment, and their knowledge of the musical genre being worked on.

18.1 Delivery Format

While the final mix should always be given to the mastering engineer at the same sample rate and bit depth as what was used during mixing, it is not uncommon to deliver a few different versions of the mix. Common versions sent are usually a mix of "Final mix," "Vox up +3dB," "Vox down −3dB," "Bass up +3dB," and "Bass down −3dB."

This way of working was fairly common until the early 2010s and allowed the mastering engineer to access slightly different mixes if he or she felt the need to. In recent times, a new type of mastering has sprung up: stem mastering. This approach fills the gap between mixing and mastering and involves processing stems rather than one stereo mix. For western acoustic ensembles, stems such as drums, tuned instruments, and vocals are quite common. In dance and electronic music, the stems required often consist of kick, drums, bass, and synths. Working this way allows the mastering engineer to have a lot of control over the balance of instruments within a mix without having to deal with large sessions. While there are mixed feelings among mastering engineers as to whether or not stem mastering should become the norm, it is more of an ethics question than a functional one. The same kind of debate regarding vocal tuning and drum sound replacement has been raging for years among mixing engineers. Stems should be delivered as interleaved stereo files rather than multiple mono to avoid issues associated with using different pan laws on the mixing and mastering systems.

18.2 Monitoring Environment

Because mastering is the last stage before a record is pressed or uploaded to digital aggregators, it is critical that the listening environment give a true representation of the song. This requirement makes the room and speakers the most important tools used during this process. For mastering in a less than ideal monitoring environment, a couple of high-quality pairs of headphones may be a viable solution. Such headphones need to be able to reproduce low end accurately, as well as give a clean midrange and top end. It is commonly accepted that bass heavy headphones such as the "Beats by DrDre" range is adequate for judging low end, while high-end open-back headphones from reputable brands will work fine for judging mids and top end.

18.3 Metering

During mastering, a few different meters will be needed. Peak (digital peaks as recognized by the DAW), RMS (average loudness similar to how the ears work), and crest factor (the difference between RMS and peak, directly related to dynamic range) meters are some that are commonly used. When you are mastering music for commercial release, the goal of the master files should be established: clean and dynamic, or loud? The dynamic range (the difference between peak and RMS level) will dictate how the master sounds. Low dynamic range (3 to 8dB) means loud music that is on par with current pop, dance, and rock music. Higher dynamic range (9 to 14 dB) means that the music "breathes" a little more, is cleaner, and has more dynamics. The current trend for loudness is roughly the following:

- 3 to 6 dB of dynamic range for pop music
- 7 to 9 dB of dynamic range for electronic dance music
- 10 to 12 dB of dynamic range for rock
- 13 to 15 dB of dynamic range for indie folk music
- 16 to 20 dB of dynamic range for jazz
- Over 20dB of dynamic range for highly dynamic music such as classical

The peak meter also needs to display inter-sample peaks (or ISPs). Two factors may prevent the DAW from registering ISPs. The first one is that there are peaks that may not be registered by the DAW as overloading the system, because the signal does not stay over long enough. Three consecutive overs is generally considered as being the norm for a clipping light to switch on, but since there may be overs that do not trigger the red light, it is important to have a dedicated meter for this purpose. The other factor is that the reconstruction filter of the digital to analog converter (DAC) being used for playback may recreate a waveform that peaks above the highest digital sample (Figure 18.1). Put simply, the highest digital sample of audio may be at −0.1 dBfs, but the process of low-pass-filtering this digital audio for playback (as a reconstruction filter does) can create a higher peak between samples.

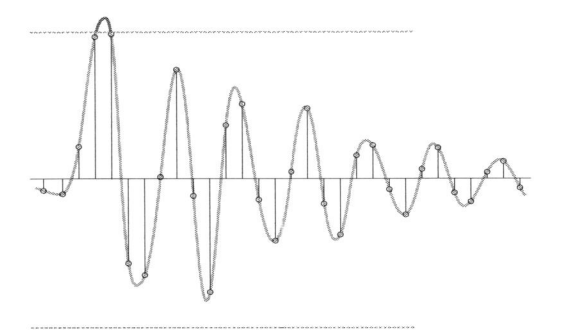

FIGURE 18.1
Inter-sample clipping.

Using an ISP meter (and acting upon the information it gives) will ultimately ensure that the music does not distort when the end user plays back audio with low-quality DACs. In order to avoid inter-sample peaks, you should first ensure that the processors you use work in oversampling mode, then run through the whole song (once the master is finished) and note where there are overs. From there, the output ceiling needs to be turned down on the final limiter and the parts that previously had overs need to be checked to ensure that ISPs have been removed there. If MP3 conversion is to happen (which is often the case for internet distribution), the MP3 file needs to be rechecked for inter-sample peaks. New overs can be created in MP3 files because of the LPF and pooling of frequencies used in the conversion. Because of this filtering of frequencies, resonance is introduced that will create higher peak level, and some frequencies that may have previously introduced phase cancellation may be removed, also resulting in higher peak level.

18.4 Room Tone

When you are mastering full albums, it is sometimes beneficial to add "room tone" between songs (Figure 18.2). This step serves the purpose of avoiding the playback going into digital silence, which may sound odd, as well as giving a feeling of glue between the different songs. If possible, room tone should be sourced in the studio where the instruments were recorded. Recording a few seconds of silence in the studio's live room is sufficient for this purpose. This "silence" can then be faded in at the end of each song and faded out at the start of the next.

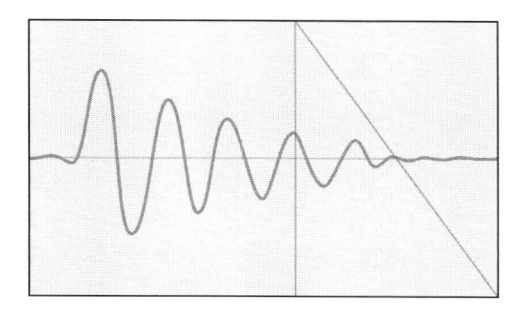

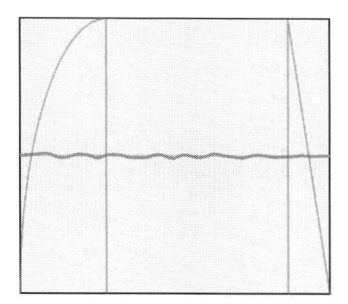

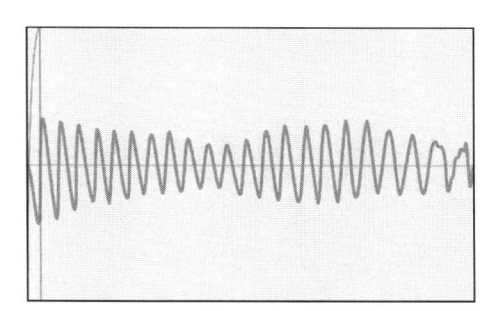

FIGURE 18.2

Room tone inserted between songs.

18.5 Track Order

While the task of picking the order of the songs on an album is not always left to the mastering engineer, there are some guidelines to follow if this task is necessary. The first is to start with one of the best songs or single to "hook" the listeners. From there, treating the album as a live performance can help keep the listeners interested and take them on a journey. Using the different aspects of each song, such as energy, pace, key, mood, and likeness, you should order the album in such a way that gently takes the listeners on ups and downs.

In some genres of music, it can also be beneficial to ensure that the key of each song is a fifth above the previous. Following the circle of fifths (Figure 18.3) when ordering songs will help keep the energy rising for the duration of the album.

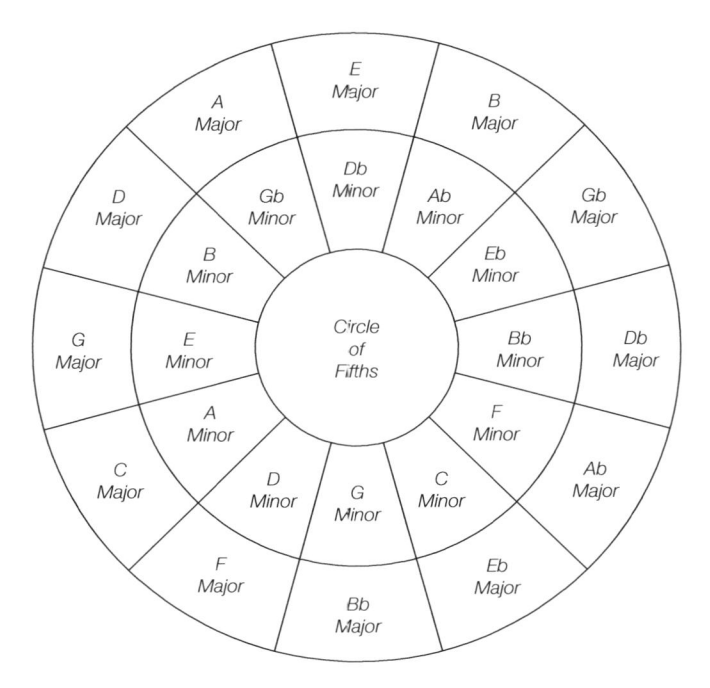

FIGURE 18.3
Circle of fifths.

18.6 Fade Out and Pause

If a CD is to be pressed for the album, a disk-description protocol (DDP) image needs to be created in an authoring software. A DDP image is the current format for submitting data to replication or duplication plants. It contains a folder with all of the songs, track data, ISRC numbers, and so on. The last creative step that you need to take into consideration at this point is the gap between tracks on an album. It is commonly accepted that a song should start at a musical value of the previous one. For example, track two may start on the downbeat two bars after the last hit of track one. This process should be done by ear by tapping your foot along at the end of the first track, and noting where it "feels" that the second track should start. This test will dictate when the first track should end, and how long the pause between songs should be. Note that this pause can be used to steer the energy levels in the right direction. For example, the gap between two high-energy songs should be shorter to keep the energy up, while the gap between high- and low-energy songs may be longer to bring the energy down before introducing the second track. This pause could even be completely removed to create a continuous album in which every track flows on from the previous one as one long song.

18.7 Mastering Your Own Mix

When you are mastering your own mix, it is preferable to mix with a limiter on the master fader rather than use a different session for mastering. Since mastering is the process of fixing mix issues, enhancing the overall sound, and increasing loudness, there is no reason to create a new session at this stage. In the case of an EP or album, tonal qualities of each song need to be matched with the next, a task that can be done only in a separate session. In this case, it is still advisable to mix through a limiter, remove it before the final export, and reapply it during the mastering session.

For more about session setup, see Box 18.1.

> **BOX 18.1**
>
> *When you are setting up a mastering session in your DAW, it is recommended that you use a "checkerboard" layout (each song on its own track, with no two songs overlapping in the timeline) to avoid having multiple songs playing at once if the solo buttons are not used properly. A small mistake such as this one can make a client present during the mastering session rather uncomfortable.*

18.8 Sending a Mix to the Mastering Engineer

Because a mastering engineer will raise the volume of the mix in the "cleanest" way possible, it is important to leave a few decibels of headroom. Around 6 dB of peak headroom is often a safe value that is required by mastering engineers.

Do not limit the master fader. Equalizing, compressing, and saturating an already limited signal introduces distortion a lot faster than doing those processes on the original signal.

No effects such as EQ, compression, stereo enhancement, and saturation should be put on the master fader unless the mix was done through these processors (i.e., they were put on the master fader before you started the mix).

Export the final mix at the same bit depth and sample rate as what the song was mixed at. There is no point in converting a mix to 32 bit/192 kHz before sending it if the session has been mixed at 16 bit/44.1 kHz. No quality will be gained in doing so.

If the mix was done with a reference track, provide it to the mastering engineer. This way he or she will know exactly what sound you are seeking. Note that the mastering engineer will be able to help only if the mix is already close to the reference. Do not expect a bad mix to be turned into a great one just by sending it to a mastering engineer!

While clear separation between low-end instruments is more of a mixing tip than a necessity, it is quite hard (although not impossible) for mastering engineers to enhance the separation between kick and bass when the low end of the song is muddy.

Bouncing stems at the same time as the main mix will allow for a quick turnaround if the stems are required by the mastering engineer. Re-exporting, uploading, and waiting for the engineer to recall the original settings and look at the project again may delay the delivery of the final master for up to a week.

Instrumental versions, live-performance versions, radio edits, a capella versions, and any other versions that need to be mastered should be asked for at the same time as the final mix is delivered. There is nothing worse for a mastering engineer than having to recall all of the hardware settings only because the artist forgot to ask for the extra versions in the first place. Engineers will most likely process the extra versions for free if they are done at the same time as the main mix.

Less is more when it comes to bass and treble in the final mix. If you are wondering whether the mix needs more bass or high frequencies, remember that mastering engineers use very high-quality EQs and other tone control devices. It is therefore preferable to let them apply these broad EQ curves with their equipment.

18.9 Mastering Tools

18.9.1 Equalizers

Equalizers can reshape the tone of a song and are therefore essential in mastering. Two different types of EQs are often used:

- Precise and clean EQ (such as the GML 9500) for removing unwanted frequencies or boosting very specific areas. Note that narrow Q settings are often used to "surgically" add or remove frequencies.
- Colored EQ (such as the Pultec EQP1-A) for adding character and gently shaping the mix. In this case wider Q settings are used.

18.9.2 Compression

The main use of compressors in mastering is to glue the elements of the mix together. Although VCA compressors are commonly accepted as the norm for this purpose because of their versatility, other types can be used, depending on the pace of the song. The gluing is achieved by ensuring that every element of the mix triggers the compressor circuit and therefore has an effect on every other instrument. For example, when the kick hits, the compressor reacts and makes the whole mix pump. When there is a loud vocal phrase, the rest of the mix is also compressed. It is this interaction that gives the gluing effect, which is also used in mixing when compressing a drums group or rhythm section, for example. The effect of a compressor inserted on groups of instruments instead of individual tracks is similar to how multichannel compression works, as opposed to dual-mono compression (see Chapter 15, "Compression, Expansion, and Gates").

Setting up a compressor for "glue" is fairly straightforward, and only a few rules apply:

- In most cases, the attack should be slow enough to let the transients through, although this attack could add punch that may not be desirable. If the aim of this compressor is also to tame dynamic range, a faster attack is necessary.
- The release should be fast enough for the compressor to release in time with the music, but not so fast that it introduces pumping or distortion. Too slow a release can "choke" the whole mix. Overall, the release should be set so that the compressor grooves with the music.
- A low ratio should be used (1.5:1 or 2:1 is usually enough). The lower the ratio, the more transparent the compression is.
- No more than 2–3 dB of gain reduction should be achieved.
- The knee setting can be used to achieve different tones: hard knee for punch, and soft knee for transparency.
- Dual-mono compressors can be useful in reducing dynamics on one side only for mixes that are constantly pushing from side to side rather than from both sides evenly. Stereo compressors can create a more even movement between speakers and help with the overall groove of the mix.

18.9.3 Reverb

When gluing cannot be achieved through compression, the use of reverb can often help in achieving a sense of "togetherness" among all the instruments of the mix. While there are no set rules as to which units to use in mastering, algorithmic reverbs are more commonly used than their convolution counterparts. The settings used will largely depend on the pace of the song, but shorter reverb times with little ERs are often a good starting point. When you are using a reverb on the whole mix, it should be mixed in at only 1% or 2% as this process can wreck the mix if overused.

The following steps can help in setting up a mastering reverb:

- Set the reverb to roughly 20% wetness.
- Play the song and run through presets, noting which ones feel the least "odd" or out of place at such high wetness percentage. You are looking for the most appropriate reverb quality here.
- Once you have found the best preset, tweak the settings to make it even less intrusive to the mix. You are working on reverb thickness and length at this point.

- Pull the wetness down to roughly 5% so the reverb is very soft.
- Tweak the settings if necessary.
- Pull the wetness down until the reverb can barely be heard (usually around 2%).

Use a high-quality reverb with a lot of settings available in mastering. This choice will ensure that it can be fine tuned to suit the song perfectly.

While this is not common practice, in the event that neither compression nor reverb helps in gluing the mix together, noise (tape hiss, vinyl hiss, and other white-noise types of generators) can be used to give the recording an overall tone and therefore masking the lack of "togetherness" among the instruments. If using noise to achieve glue, ensure that it is constant throughout the length of the song. If the noise must be turned off during certain sections, fade it in and out over a long period of time to ensure it does not attract the attention of the listener.

18.9.4 Parallel Compression

For overly dynamic mixes, parallel compression may be required. This process will allow for the song to retain its original dynamics while the compression raises the overall loudness. This technique should be used only with mixes that have not been compressed a lot, as parallel compression works only if the original signal contains enough dynamics. In most cases, a mix of regular downward compression (in the way of the glue compressor previously mentioned), and parallel (or upward) compression can tame the dynamics of the mix by turning down the loudest peaks and turning up the softest parts of the signal (Figure 18.4).

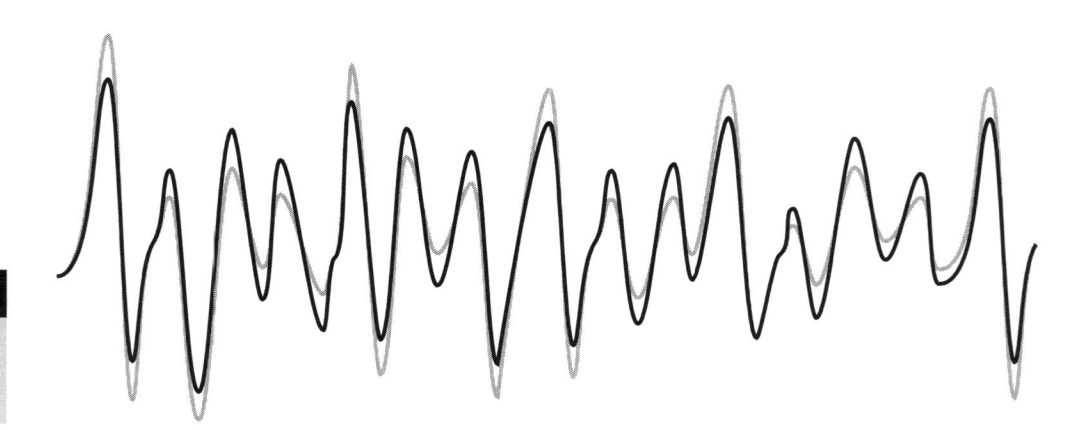

FIGURE 18.4

Downward and parallel compression to reduce dynamics.

18.9.5 Dynamic Equalizers

There are two main uses for multiband compressors and dynamic equalizers in the mastering process: as a frequency-dependent dynamics control device and as an EQ which turns on and off dynamically. When you are using this tool as a dynamic EQ, it often acts just as a split band de-esser would, to remove unwanted frequencies but only when they are too prominent, such as the harshness associated with loud hi-hats. It can also be used to raise or lower the volume of a particular frequency dynamically. For example, a kick drum can be raised in volume without affecting the bass guitar by ensuring that only the kick drum triggers the volume boost. It can be useful to think of dynamic EQs as "fixing" processors.

18.9.6 Multiband Compression

When used as a dynamic control and gluing device, the multiband compressor is aimed on polishing the mix by using three bands: lows, mids, highs. In this case, it is important to understand that the multiband compressor is acting in the same way as the previously mentioned glue compressor, but working on separate bands, giving finer control over the overall sound. The following steps are useful guidelines for using a multiband compressor for polish:

- The detection circuit should work on the average level (RMS) rather than peak level to allow for smoother compression.
- The crossover for the low end should be placed where only the subs (kick and bass) can be heard, usually around 150 Hz–200 Hz. The split between the mids and highs should be done at the point where the vocals lose intelligibility, usually around 4 kHz–6 kHz. Start with steep cutoff slopes for the crossover filters, and once they have been set, experiment with gentler slopes.
- The next step is to treat the compressor as if it were a simple wide-band processor and link all three of the bands to set the attack and release settings. These settings should remain the same for all bands, as different values could give the impression that the mix is "moving" at different paces, depending on frequency content. Slower attack settings will allow the transients to go through unaffected, which is often desired, as the limiter that follows this compressor will deal with them. The release should be set to work in time with the song's pace.
- Once the attack and release have been set for all bands, set the thresholds so that every band has around 2–3 dB of gain reduction.
- Next comes the ratio and knee settings. Use a hard knee for punch and soft knee for smoothing out transients. Start with a high ratio (2.5:1) and hard knee for the subs, medium ratio (2:1) and medium knee for the mids, and low ratio (1.5:1) and soft knee for the highs, then tweak if necessary. This approach ensures

that the high end is not too punchy and piercing, while you keep the low end hard hitting.
- If the compressor allows parallel compression (via a dry/wet knob) on a band by band basis, experiment with this setting to see if bringing back some of the original uncompressed signal is beneficial.
- Once the compressor has been set, do not forget to match the output with the input level. Remember that the overall aim of this compressor is to give a final polish to the song by controlling the different bands separately.

18.9.7 Limiting

Brick-wall limiters are used to obtain louder mixes in the second last process of the mastering chain. Limiters are compressors with very fast attack, hard knee, high ratio, and auto-gain. Some have more controls than others, but even the simplest ones will have a threshold (which controls both the threshold at which the limiter kicks in and the makeup gain) and output ceiling (which dictates the maximum peak value that the output of the limiter can reach). Most will also have a release setting, which can be used to get rid of distortion (with slower release times) if the limiter is working too hard. It is often advised to start at rather fast settings, and slow the release down to remove distortion. Care should be taken with this control as it can change the balance of transient instruments compared with sustaining instruments within the mix.

The general rule when using limiters for mastering is that they should be catching only the highest peaks and reducing them by a maximum of 3–4dB. This rule allows for "clean" limiting and does not introduce any distortion. When mastering for loudness, some engineers use more than one limiter in series, often in a combination of general and multiband limiting. Note that a good starting point for the ceiling control is −0.3 dBfs, as it provides a little "padding" for ISPs and is considered to be the least noticeable difference in level. In any case, using an ISP meter as previously mentioned is the only true way of ensuring that this control is properly set.

For more about loudness, see Box 18.2.

BOX 18.2

A mastering trick used to achieve extra loudness is to overload the input of a high-quality ADC. When doing so, you must be careful to overload the converter for only a very short time. This technique should be used only with very high-quality converters.

18.9.8 Saturation via Transformers, Tubes, and Tape

Since the advent of digital audio, the accidental use of saturation through outboard equipment and tape has been declining. Plug-in manufacturers have recognized that there is a need for emulations of hardware and their nonlinearities. While distortion can be used to enhance "cold" digital audio, it is better to use this process during the mixing stage, as the nonlinearities can help achieve both separation and glue between instruments. In a mastering situation, a simple tape emulation can help warm up a mix, but you should be careful when using more than one stage of digital saturation. If running through outboard equipment, slight saturation will be introduced in a more "musical" way and will achieve better results.

18.9.9 Dither

Dither is the process of adding low-level noise to replace the distortion introduced when you are reducing bit depth in fixed-point processing. This distortion is caused by quantization errors from truncating word lengths. When you are using dither, the low-level information represented by the least significant bits is randomized to allow for parts of it to be recreated in a lower bit-depth file. This technique effectively fools the ear into hearing a higher bit depth than what is available, a concept more easily understood if you think of pictures. When you are visualizing a grayscale picture that needs to be converted into black and white (Figure 18.5), one option would be to round off every gray pixel to its closest black or white value. This step would be the equivalent of truncation in audio (not using dither) (Figure 18.6). Another option would be to randomize values that are close to the halfway point between black or white to effectively "blur" the color switch area, fooling the eye into thinking that the picture contains more colors (Figure 18.7).

FIGURE 18.5
Original image.

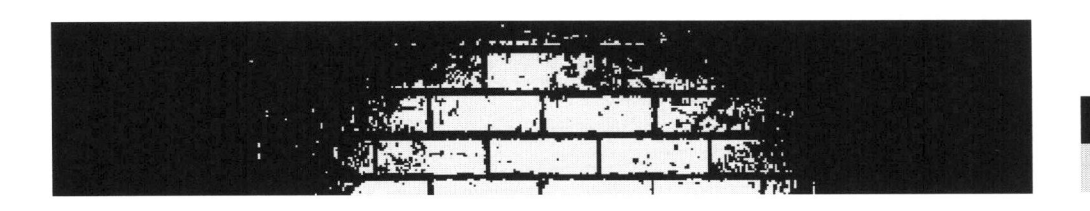

FIGURE 18.6
Truncated image.

FIGURE 18.7 Dithered image.

Dither happens "under the hood" when you are working with some third-party plug-in and certain types of tracks in modern DAWs. For example, Pro Tools used to give the user a choice of working with a dithered or regular mixer by selecting a specific plugin in the system's plugin folder. In terms of mastering, dither is the very last process to be done before final export. It is important that no other digital process such as sample-rate conversion, or further digital effects are applied after the dither is inserted, as the result would be the quantization errors remaining as well as the noise being added. For most applications, a shaped dither will be more appropriate. Shaping allows for the dither noise to be placed in the frequency spectrum where it is least noticeable. If the song is to be altered after the master is released (as often happens in the way of unofficial remixes and edits, for example), using a triangular probability density function (TPDF) dither tends to be the better option, as the noise is evenly spread across the spectrum. If you are re-dithering by using shaped dithers, the whole song could sound thin, as too much noise will be added in the high end of the spectrum. Overall, re-dithering is mostly an issue when you are working at 16 bit, because multiple dithering can become quite audible. At 24 bit, multiple dithering is less of an issue; the noise introduced is lower than the DAC's noise floor (around 21 bit).

To judge dither quality, lower the gain of the music significantly (around −40 dB should be enough) and listen through high-quality headphones. Note the tone and distortion introduced by each dither type and choose the most appropriate for the song. Be careful when you are using different shapes, as they can be problematic to some listeners, depending on the quality of their hearing at different frequencies. For example, dithers containing a lot of high-frequency content may not be the best suited to children's music, because of their hearing sensibility in this area of the spectrum.

Dither is most useful in "clean" genres of music such as jazz, classical, or folk, and can often be omitted if the master already contains distortion as part of the music itself. Thus most modern music can be freely altered without the use of dither. It is also unnecessary to dither when you are converting from a floating-point bit depth such as the common 32-bit float format available in most DAWs, since the least significant bits being dithered are not part of the mantissa (the bits that will remain after the bit-depth reduction).

18.9.10 Mid-Side Processing

Mid-side decoding can be used to access individual parts from a stereo file. Just as EQs can be used to separate instruments on the basis of their frequency content,

MS can be used to separate sounds depending on their phase relationship. For example, if the lead vocals need de-essing, there is no need to work on the whole mix. De-essing the center information will ensure that the information to the sides stay intact. In much the same way, if a track contains too much reverb, processing the sides on their own will be more transparent than working on the whole mix. Note that a mono instrument panned to one side will still end up in the middle channel. Only stereo sounds that have different frequency content in either of the sides (and therefore somewhat out of phase) will end up in the sides channel.

There are a few different ways to decode a signal into MS, each having pros and cons. The first technique offers the benefit of not having to decode the signal back from MS to LR. Simply adding the three stems (middle, left side, right side) together plays back properly as LR (Figure 18.8). The downside to using this technique is that three stems are created, and thus using stereo outboard equipment is limited.

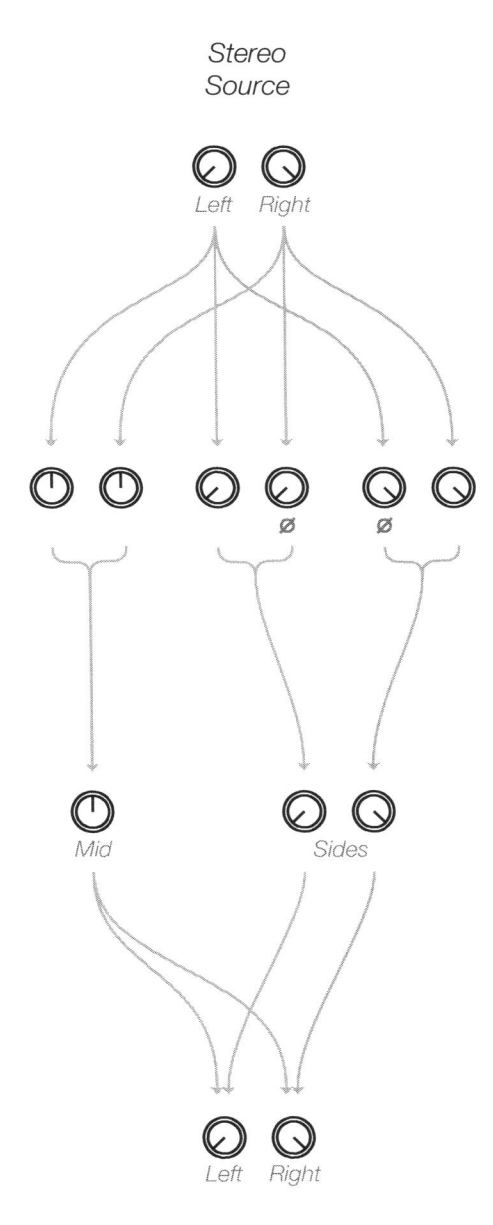

FIGURE 18.8

Mid-side decoding technique #1.

AUDIO PRODUCTION PRINCIPLES

Both the second and third techniques offer only two stems (middle and sides), which makes them best suited to using outboard processors. The small downside to these techniques is that the signal must be decoded from MS to LR before the signal can be played properly (if this is not done, the middle channel will play back in the left speaker, while the sides channel will play back in the right speaker). Both techniques two and three process the signal the same (Figure 18.9). The second technique, while simpler to set up, requires a specialized plug-in to be inserted on the tracks. The third technique takes longer to set up, but can be done natively in every DAW.

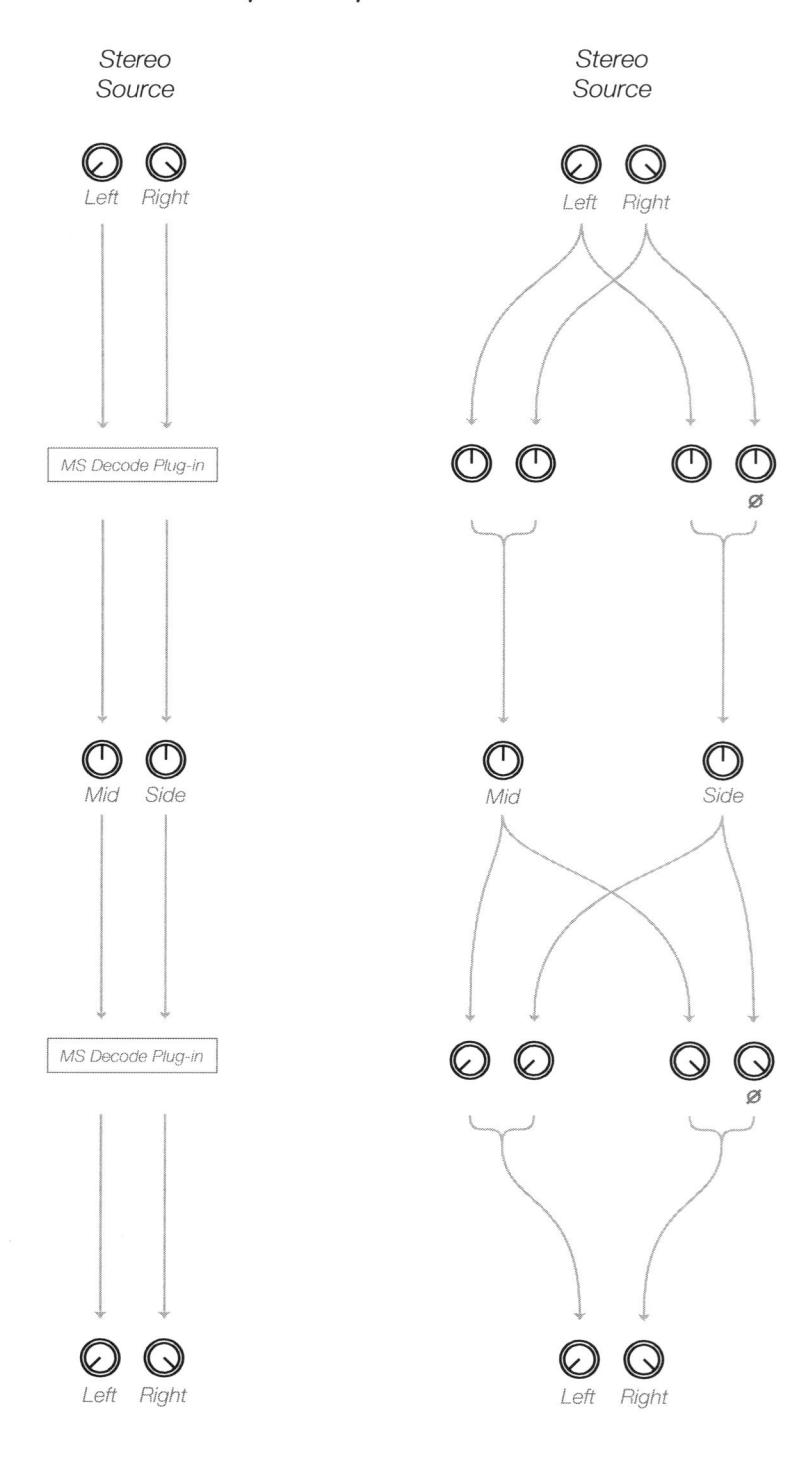

FIGURE 18.9
Mid-side decoding technique #2 and #3.

MS processing can also be used to alter the stereo width of the whole song. This method is often quite useful, as MS processing does not affect mono compatibility. For example, lowering the volume of the center channel will make the song sound wider. More advanced uses of MS processing to alter the stereo width include the following:

- Equalizing the low mids up on the sides (using a low shelf) can increase the sense of width (Audio 18.1 ▶).
- Using slight saturation on the sides can make the song wider (Audio 18.2 ▶). This method works especially well when you are processing the low mids (150 Hz–400 Hz) with multiband saturation.
- One of the best ways to ensure that the song is consistently wide is to parallel-compress the low mids on the sides only (Audio 18.3 ▶). This step ensures that the low mids are always present on the sides and thus in turn creates a more even width throughout the song.

As mentioned in Chapter 12, "Mixing," MS processing should also be used to ensure that the tonality of the mix is similar when played in mono and stereo. For example, if the whole mix is dull when played in mono, brightening the center and making the sides duller will make the mono and stereo mixes closer to each other in terms of tone. Similar processing is often necessary with reverb, EQ, and saturation.

For a tip about outboard equipment level matching, see Box 18.3.

BOX 18.3

If you are using non-stepped outboard equipment, running the audio in MS instead of LR through the equipment can ensure that the relative level of both left and right channels remain intact. Instead, if the dials are not exactly in the same position, only the overall width of the sound will be affected.

Note that in order to match the mid and sides tones, it is also possible to "re-inject" the content of one into the other. Using the previously mentioned second and third techniques to decode a stereo signal into MS, you can match tones simply by sending a copy of the sides channel into the center (the most common use of this technique) or sending a copy of the center channel into the sides (the less common use). The result of sending a copy of the sides into the center channel is that the reverb and other bright mix elements will now be present when the mix is played back in mono. Be careful here, as this step can make the mix narrower, since there is more common information in both channels. In order to keep the original stereo width, but still "re-inject" the sides into the center, you will find it useful to add a short delay of up to 20 ms to this copy (Figure 18.10).

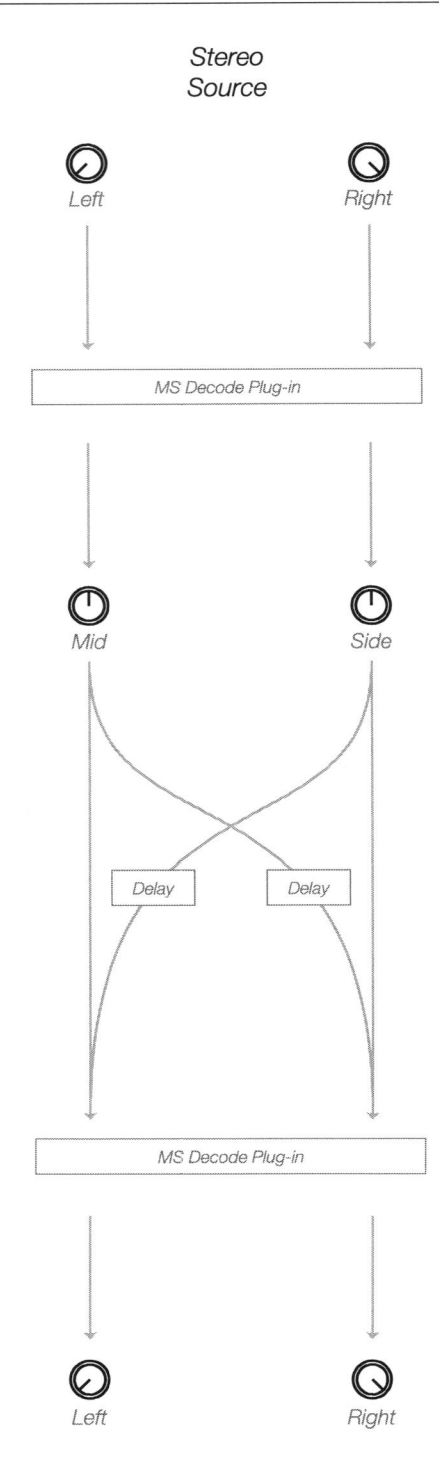

18.10 Deliverables

A mastering engineer will need to deliver the final files in a variety of formats, depending on the intended release medium. While "Red Book" CD format has been the norm since the 1980s, only some of its specifications still apply to digital releases. The current norm is for the final file to be 16-bit/44.1 kHz uncompressed audio (1411 kbps). With advances in technology, cheap hard disk space, and Internet bandwidth, this norm is likely to shift to higher-quality audio. Because of this likelihood, it is recommended that you keep a copy of the master in 24-bit/96 kHz format.

In the late 2000s, iTunes introduced a specific way of encoding audio into an AAC file (the format that Apple currently uses for music release) called "Mastered for iTunes." While the encoding is a lossy process and therefore sounds inferior to the original high-quality file, using a high-quality 24-bit/96 kHz file as the source material will ensure the best results are achieved.

In an effort to end the "loudness war," web-based music distribution sites have introduced loudness normalization in the late 2010s. While each distributor has implemented this concept slightly differently, the overall concept is the same: similar to how peak normalization looks at the loudest peak in the wave file and brings it to a specific value (bringing the rest of the peaks up and down as a result), loudness normalization makes every song as loud one another by analyzing them and normalizing the result to a set

FIGURE 18.10

"Re-injecting" center into sides and sides into center by using MS decoding technique #2.

value through a change of volume. The result of this process means that highly limited masters (such as in pop music) sound as loud as more conservative masters when played back through these online distributors. A flow-on effect of this process is that songs that contain more "life" and "breathe" more sound better than their highly-limited counterparts. This is a consideration which must also be taken into account when deciding of the final level for a master.

18.11 MP3

The MP3 is the most widely used digital audio format and should therefore be understood by all engineers in order to make informed decisions during the mixing/mastering stages of a project. Even though some may argue that MP3 is no longer required now that internet speeds have gone up (MP3 really took off during a time where dial-up internet was standard), this format is still very much used all throughout the world.

It is important to understand that MP3 is a lossy format and a loss of information therefore occurs when a song is encoded. MP3 groups audio frequencies into different bands and discards sounds when their volume drops below a certain threshold (when compared to adjacent frequency information). This factor alone means that at lower bit rates, the more "pooling" of frequencies occurs, the more separation between instruments is reduced.

MP3 also discards more and more high-frequency content in audio as lower bit rates are used. The results are a loss of clarity (in terms of high-frequency definition), transient definition (transient information of drums mostly consists of high frequencies), reverb (which is often lower in level than dry signals), and stereo information (high-frequency information is panned harder than low frequency).

Simply put, MP3 distributes a number of available bits to encode audio. How those bits are distributed depends on the options selected before encoding a file. The way that those bits are distributed can greatly affect the quality of the final product, even at the same file size.

Constant bit rate (or CBR) will distribute bits evenly over time: 192 kbps will use exactly 192,000 bits per second, giving a predictable final file size (good when an exact file size must be obtained).

Variable bit rate (or VBR) will distribute more bits to parts of the song that are more complex in terms of frequency content, and fewer bits to parts of the song that are simpler. Thus breakdowns of a song may be "given" fewer bits compared to choruses, for example. This way of encoding focuses on achieving a particular output quality, regardless of file size.

Average bit rate (or ABR) stands halfway between CBR and VBR. The bits distributed will vary, depending on the content of the audio being encoded, but they will hover around a set bit rate to achieve a more predictable file size.

Stereo encoding means that the MP3 file will be encoded as two mono files, just as a regular uncompressed audio file would.

Joint stereo encoding converts the left and right channels into MS channels for the encoding. Since the side channel contains less information (especially in the low end), this method of encoding can give better results at lower bit rates.

If the highest-quality MP3 is to be produced, "320kbps, CBR, stereo" encoding will produce the best results. As the quality is lowered to reduce file size, switching to ABR or VBR joint stereo will give better results for smaller file sizes.

Chapter **19**

Conclusion

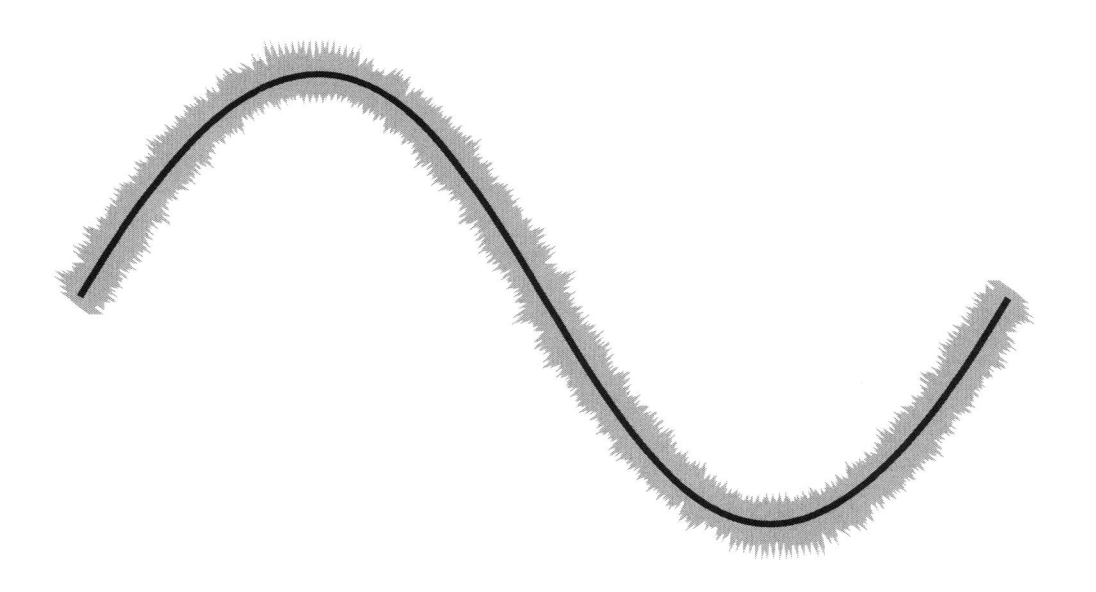

This book has touched on many subjects related to the production and engineering of music. Whether you feel that the technical elements or the more abstract concepts such as workflows and mindsets play a larger role in achieving great-sounding records, be sure to remember the most important aspect of this craft: to create emotions through music and sound. Since ancient times, and long before recording technology ever appeared, people have been producing sounds for this purpose. So the next time you are faced with a problem in the studio, always remember the end goal. Never forget to question what you know and do not be afraid to break the rules; this boldness is how greatness and originality is created. Coming back to what has been mentioned in the introduction of this book, I emphasize that a good balance of experience, knowledge, and equipment is necessary to produce great records. It is often said that at the start of your journey, knowledge will be the most important aspect to develop. After a while, your knowledge will uncover what is lacking in your equipment, which is an easy fix. What takes time and patience is the building of experience. No one becomes a superstar producer or engineer overnight, and the key is to keep working until you build experience. Never forget the "10,000-hour rule" necessary to become an expert in your field and do not be disappointed by your lack of competences, it will take hundreds of records before you can consistently achieve great sounds!

So, where to now? Keep learning, forever. Learn by reading books and manuals. Learn by watching video tutorials and how fellow engineers work. Learn by analyzing your own productions and teaching concepts to others. Listen to music and learn to appreciate good productions, no matter the genre. Learn by transferring concepts from different disciplines to sound production. It is amazing how a seemingly entirely different field such as graphic design or film production can spark ideas for new techniques in audio production. Your capacity to achieve greatness is limited only by your own mind.

Good luck in your quest for great sounding records, and above all, have fun!

Index

Note
Italic page numbers indicate figures or tables

16–bit. *See* bit depth
24–bit. *See* bit depth
32–bit float. *See* bit depth

A
AAC (Advanced Audio Coding), 246
AB microphone technique. *See* spaced pair microphone technique
ABR. *See* MP3
absorption. *See under* acoustic treatment
acoustic guitar, 99
   creating interest with, 104
   frequency content, *167*
   mixing, 103
   recording, 99–103
acoustic panels. *See* acoustic treatment
acoustic treatment, 12–17, 19. *See also* recording room
   absorption, 16–17
   bass trapping, 14–15
   diffusion, 16–17
   live end/dead end (LEDE), 17
   placement of, 12–17
   reflection-free zone (RFZ), 12–13
active EQ, 167
aliasing, 47–48, 180
all-pass filters, 124. *See also* linear phase; filters
analog emulation, 149, 225, 227
analog to digital converter (ADC), 45–48. *See also* digital to analog converter (DAC)
   overloading of, 240
attack of waveform. *See* transients
attack time of compressors, 175–177, 180–182, 185–190, 202, 237, 239. *See also* limiters
attack time of expanders/gates, 204–206
automation. *See also* micro-modulations
   as a creative tool, 51, 104, 133, 141–142, 144–145
   for dynamic range control, 85–86, 110–111, 137, 189

B
balance
   drum kit, 74
   frequency, 138–140, 143–144, 155–157, 222
   mix, 137–140, 143–145
band-pass filters, 123
bandwidth (Q). *See under* EQ
bass drum. *See* kick drum
bass frequencies, 8, 31–32, 35, *138*, 146, *167*. *See also* proximity effect; monitors: subwoofer
bass guitar, 82
   creating interest with, 87
   frequency content, *167*
   mixing, 84–87, 129
   recording, 82–84
bass trapping. *See under* acoustic treatment
bidirectional microphones. *See also* microphones; polar patterns
   attributes, 31, *35*, 108
   recording techniques, 38–42, 66–67, 100–101, 103
bit depth, 45–47, 242. *See also* dither
bit rate. *See* MP3
Blumlein microphone technique, 38–42, 67, 101
brick-wall limiters. *See* limiters
bus compression, 50, 144, 148, 184, 186. *See also* coherence; glue: compression

C
cardioid microphones. *See also* microphones; polar patterns
   attributes, 30, *35*, 108
   recording techniques, 37–42, 64–69, 100–102
CBR. *See* MP3
CD (compact disc), 234, 246
chorus effect, 223. *See also* delays
   uses of, 87, 143, 214, 219
clipping, 46, 78, 92, 103, 178, 226–227, 231–232. *See also* distortion; gain staging
close microphones, 33, *35*, *42*, 69, 74
coherence, 144–145. *See also* glue: compression; reverberation
coincident microphones, 37–39, *42*, 66, 100–101, 183. *See also* Blumlein microphone technique; mid-side microphone technique; XY microphone technique

Index

comb filters, 124, 222–223
compression, 174–175, 236. *See also* limiters
    acoustic guitar, 103
    bass, 85–86
    drums, 64, 75, 78
    frequency specific, 199–202, 238–240
    hardware, 175–178
    mixing with, 138–139, 144
    setup, 185–188, 237
    settings, 179–184
    techniques, 171, 188–199, 202–203, 218–220, 222, 238–240
    voice, 110–111, 117–118
condenser microphones, 30, 32, *35*
control room, 6–7. *See also* acoustic treatment; monitors
convolution reverb, 211–212, 217. *See also* reverberation
crest factor. *See* metering
critical listening, 2–4, 134, 167
crossfade, 114–115
crossover. *See* monitors: subwoofer

D

damping a snare drum, 60
DAW, 22–23, 45, 154. *See also* Pro Tools
de-esser, 75, 117, 199–201. *See also* multiband compressors
delays, 118–119, 219, 221–224
    further processing of, 118, 196
    panning using, 153, 156
deliverables, 246–247
depth of field, 12, 132, 140, 159–160, 169, 215–216
DI box, 83
diffusion. *See under* acoustic treatment
digital to analog converter (DAC), 232. *See also* analog to digital converter (ADC)
DIN. *See* near-coincident microphone techniques
diode bridge compressors. *See* compression: hardware
distortion, 92, 119, 128, 179, 206, 226–228, 240–241. *See also* clipping
dither, 241–242
doubler effect, 223
drum kit, 58–60
    creating interest with, 79–80
    frequency content, *167*
    mixing, 71–79
    recording, 60–71
    skins, 58–59
dynamic microphones, 30–32, *35*
dynamic range, 135, 148. *See also* bit depth; gain staging
    of analog equipment, 45–46, 90, 135

E

ear. *See* localization of sound source
echo chambers, 209
editing drums, 71–73
editing voice, 114–116
electric guitar, 89–93
    creating interest with, 97
    frequency content, *167*
    recording, 93–96
envelope, 3, 125
EQ, 161–163, 165–167, 236. *See also* filters
    bandwidth (Q), 163–164
    drums, 74, 77
    electric guitars, 103
    mixing, 138–142, 168–173, 190, 217–218
    monitoring system, 19
    panning, 152, 155
    voice, 110–111, 117
expanders, 76, 203–207

F

FET compressors. *See* compression: hardware
filters, 11, 123–124, 135–136, 164–165, 222
flam effect, 223
flanger effect, 223
frequency chart, *169*

G

gain staging, 45, 135, 143
gates, 76, 203–207
glue. *See also* coherence
    compression, 148–149, 186, 236–237, 239–240
    noise, 232
    reverb, 212, 237–238
graphic equalizer, 173
guitar, acoustic. *See* acoustic guitar
guitar, electric. *See* electric guitar

H

hard clipping, 226. *See also* clipping
headphones. *See under* monitors
high-level compression, 202. *See also* compression
high-pass filters, 123, 135–136

I

impulse response. *See* convolution reverb
inter-sample peaks (ISP), 231–232. *See also* gain staging; metering

J

jitter, 47

K

kHz. *See* EQ; sample rate
kick drum, 58–60, 75, 78, 85
knee, 181

Index

L
large-diaphragm microphones, 31–32, 109
levels. *See* balance
LFO, 125
limiters, 177, 226, 240
linear phase, 165–166
listening. *See* critical listening
listening environment. *See* control room
live end/dead end (LEDE). *See under* acoustic treatment
live recording, 48–51
localization of sound source, 40–41
look-ahead, 205
lossy audio, 246–247. *See also* MP3
loudness, 186–187, 240, 246. *See also* limiters; metering
low-level compression, 117, 188–189, 202
low-level monitoring, 146–147
low-pass filters, 91, 123, 135–136, 164–165

M
mastering, 148–149, 229–230, 234–236
metering, 46, 136–137, 158, 231–232
micro-modulations, 79–80, 87, 97, 133, 143, 145
microphones, 28–32
    ambience, 63–64
MIDI, 87, 126, 129
mid-side microphone technique, 38–39, *42*, 65, 100
mid-side processing, 137, 140, 156–157, 184, 242–246
minimum phase, 166
mixing. *See* balance; workflow
monitors
    headphones, 54–55, 231
    main monitors, 5–9
    subwoofer, 9–12
mono compatibility, 136–138, 140
MP 3, 247–248
multiband compressors, 75, 118, 171, 199–202, 239–240
multi-tap delay, 224

N
near-coincident microphone techniques, 39–40, *42*, 67–68
normalization, 75–76, 246
NOS. *See* near-coincident microphone techniques
notch filters, 123

O
omnidirectional microphones. *See also* microphones; polar patterns
    attributes, 31, *35*, 109
    recording techniques, 35–36, 39–42, 63–65, 100–102
optical compressors. *See* compression: hardware
order of processing when recording, 51, 55, 61, 95
order of processing when mixing, 110, 139, 187–188, 237–238, 239–240
ORTF. *See* near-coincident microphone techniques
oscillators. *See* waveforms
oversampling, 48, 180, 232. *See also* sample rate

P
panning, 137, 151–154, 157
    instruments, 74, 116, 130, 158–159
parallel compression, 117, 188–189, 202, 207, 238
passive EQ, 167
patching list, 23, 60–61
phase alignment, 73–74, 136–137
phaser effect, 223
ping-pong delay, 224
plate reverb, 209
playback system. *See* monitors
polar patterns, 29–31, *35*, *42*
pop filter, 109
pre-delay, 219
preproduction, 20–22
Pro Tools, 25, 47, 114
proximity effect, 8, 30–31, 34–35, 89
PWM compressors. *See* compression: hardware

Q
quantizing, 71–73, 84–85

R
RAI. *See* near-coincident microphone techniques
ratio setting of compressors, 180
recording room, 58, 82, 89, 99, 106. *See also* reverberation
    effect on microphone selection, 29–33, *35*, 100
    effect on recording studio selection, 49
    effect on stereo recording technique selection, 38–40, *42*, 67–68, 101
    enhancing natural reverb in a recording, 190, 202
    room tone in mastering, 232
release times of compressors, 175–177, 181, 185–190, 202, 237, 239
reverberation, 140, 144–145, 208–217, 237–238
ribbon microphones, 30, *35*
room-analysis software, 17–19
room-correction software, 19
RT 60, 16

S
sample rate, 47–48. *See also* aliasing; oversampling
saturation. *See* distortion

Index

short delays, 153, 222–223
sibilance, 109–110, 114–115, 199–200
side-chain processing, 118, 171, 190–199, 219
slapback delays, 118–119, 224
small-diaphragm microphones, 31–32
snare drum, 59–60, 62
spaced pair microphone technique, 35–36, *42*, 64, 101–102
speakers. *See* monitors
spring reverb, 210
stereo microphone techniques, 35–42, 64–69, 99–102
subwoofer. *See under* monitors
synthesis, 120–121, 128

T

talkback, 54
tape, 45, 178–179. *See also* compression
tape delay, 224
tempo, 23–24
three-to-one rule, 36–37, 41
threshold control of compressors, 179, 207
time stretching, 72–73
transients. *See also* attack time of compressors
    drum kit, 59–60, 62
    effect of distance on, 32–33
    effect of recording room on, 58
    electric guitar, 91
    monitoring fidelity, 12, 146

MP 3, 247
    processors affecting, 75–76, 78, 117, 166–167, 186, 188–189, 206, 219
    response of microphones, 29–31, 34–35
synthesisizer, 125
transient designers, 186
tremolo, 126, 128
tuning drums, 58–59

V

vari-mu compressors. *See* compression: hardware
VBR. *See* MP3
VCA compressors. *See* compression: hardware
vibrato, 86, 127, 128
voice, 106–107
    frequency content, *167*
    mixing, 114–119
    recording, 107–114

W

waveforms, 121–122
workflow, 133–146

X

XY microphone technique, 37, *42*, 65–66, 101

Z

zero crossing, 115

Printed in Great Britain
by Amazon